LEVERAGING AMERICA'S AIRCRAFT CARRIER CAPABILITIES

Exploring New Combat and Noncombat
Roles and Missions for the U.S. Carrier Fleet

John Gordon IV, Peter A. Wilson, John Birkler,
Steven Boraz, Gordon T. Lee

Prepared for the
United States Navy

Approved for public release;
distribution unlimited

NATIONAL DEFENSE
RESEARCH INSTITUTE

The research described in this report was prepared for the United States Navy. The research was conducted in the RAND National Defense Research Institute, a federally funded research and development center sponsored by the Office of the Secretary of Defense, the Joint Staff, the Unified Combatant Commands, the Department of the Navy, the Marine Corps, the defense agencies, and the defense Intelligence Community under Contract DASW01-01-C-0004.

Library of Congress Cataloging-in-Publication Data

Gordon, John, 1956–
    Leveraging America's aircraft carrier capabilities : exploring new combat and
noncombat roles and missions for the U.S. carrier fleet / John Gordon IV,
Peter A. Wilson, [et al.].
        p. cm.
    "MG-448."
    Includes bibliographical references.
    ISBN 0-8330-3922-9 (pbk. : alk. paper)
        1. Aircraft carriers—United States. 2. United States. Navy—Aviation. 3. Naval
strategy. 4. United States—Military policy. I. Wilson, Peter (Peter A.), 1943–
II. Title.

V874.3.G67 2006
359.9'48350973—dc22

                                                                                  2006008529

The RAND Corporation is a nonprofit research organization providing objective analysis and effective solutions that address the challenges facing the public and private sectors around the world. RAND's publications do not necessarily reflect the opinions of its research clients and sponsors.

RAND® is a registered trademark.

Cover design by Stephen Bloodsworth

© Copyright 2006 RAND Corporation

All rights reserved. No part of this book may be reproduced in any form by any electronic or mechanical means (including photocopying, recording, or information storage and retrieval) without permission in writing from RAND.

Published 2006 by the RAND Corporation
1776 Main Street, P.O. Box 2138, Santa Monica, CA 90407-2138
1200 South Hayes Street, Arlington, VA 22202-5050
4570 Fifth Avenue, Suite 600, Pittsburgh, PA 15213-2612
RAND URL: http://www.rand.org/
To order RAND documents or to obtain additional information, contact
Distribution Services: Telephone: (310) 451-7002;
Fax: (310) 451-6915; Email: order@rand.org

Preface

On numerous occasions over the past 50 years, U.S. military and civilian defense leaders have relied on aircraft carriers and their air assets, not only as key forward-based elements of the nation's deterrent and warfighting force but also when the United States has needed to project military power, engage in hostile operations, provide humanitarian relief, or fulfill a range of other hostile and nonhostile missions. Because they offer unparalleled mobility, provide sustained military presence, can send signals of U.S. concern and possible actions, and free the United States from having to conduct flight operations from foreign bases or obtain permission from foreign powers to fly over territory, aircraft carriers likely will continue to be an asset of choice for years to come. Indeed, it is entirely possible that, as the United States seeks ways to stretch its defense dollars, pursue the Global War on Terrorism, and meet other national-security challenges, policymakers will increase their reliance on aircraft carriers, using them more often and in more situations than they have in the past, especially if the vessels have the additional capabilities to respond appropriately.

The current and expected use of aircraft carriers led the U.S. Navy in fall 2004 to commission the RAND Corporation to explore new and nontraditional ways that the United States might be able to employ aircraft carriers in pursuit of traditional and emerging military and homeland defense missions. Over six months, RAND created and convened two Concept Options Groups (COGs)—small groups of experienced military and civilian experts, defense analysts, and potential users who work together to identify promising ways to employ military might in nontraditional ways—to explore possible nontraditional roles for aircraft carriers. One COG explored and identified new ways that aircraft carriers could be used in combat operations; the second COG examined ways that the vessels could be used in noncombat, homeland security missions or to help the nation recover from terrorist attacks or natural disasters in U.S. territories.

This monograph summarizes the activities, findings, and recommendations of both carrier COGs. It should be of special interest to the Navy and to uniformed and civilian decisionmakers with responsibilities related to naval and carrier operations, maritime domain awareness, or homeland security.

This research was sponsored by the Program Executive Office–Aircraft Carriers, Naval Sea Systems Command, and was conducted jointly within the Acquisition and Technology Policy Center and the International Security and Defense Policy Center of the RAND National Defense Research Institute, a federally funded research and development center sponsored by the Office of the Secretary of Defense, the Joint Staff, the Unified Combatant Commands, the Department of the Navy, including the Marine Corps, the defense agencies, and the defense Intelligence Community.

For more information on RAND's Acquisition and Technology Policy Center, contact the Director, Philip Antón (email: Philip_Anton@rand.org; phone: 310-393-0411, extension 7798; mail: RAND, 1776 Main Street, Santa Monica, CA 90407-2138).

For more information on RAND's International Security and Defense Policy Center, contact the Director, James Dobbins (email: James_Dobbins@rand.org; phone: 310-393-0411, extension 5134; mail: RAND, 1200 Hayes Street, Arlington, VA 22202-5050). More information about RAND is available at www.rand.org.

Contents

Figures

Table

Summary

To meet combat and noncombat demands in the future, the United States' aircraft carriers will require a range of capabilities that they do not currently possess. Carriers will need to be better able to mix and match personnel, aircraft, and other assets to emerging and evolving tasks. They will need to perform more-extensive surveillance and reconnaissance, conduct air operations at greater distances, and be equipped to operate in nuclear environments. And they will need to be more modular, deploy on shorter notice, and be prepared to handle more casualties than they can today.

So concludes this analysis that RAND conducted in 2004 and 2005 on behalf of the U.S. Navy. Over six months, RAND created and convened two small groups of experienced military and civilian experts, defense analysts, and potential users to investigate possible nontraditional roles for aircraft carriers. Nontraditional uses of aircraft carriers include, for example, carriers being used by aircraft of the Army or Air Force and new or different mixes of capabilities being brought aboard the ships. One group explored and identified new ways that aircraft carriers could be used in combat operations; the other examined ways that the vessels could be used in noncombat homeland-security missions or to help the nation recover from terrorist attacks or natural disasters.

These groups explored two fundamental questions: How have aircraft carriers been used in nontraditional ways in the past? What nontraditional roles and missions might aircraft carriers be asked to perform in the future? They addressed these questions by cataloging how and under what conditions aircraft carriers have been employed in the past and by identifying circumstances that the United States might encounter in the next 20 or 30 years that could require aircraft carriers to be employed in nontraditional roles. The analysis also examined alternative ways that carriers could be properly equipped with an appropriate mix of capabilities for those roles.

New or Nontraditional Roles for Aircraft Carriers?

For more than seven decades—in circumstances stretching from before World War II through the Global War on Terrorism to the 2004 Southeast Asia tsunami—aircraft carriers and their embarked air wings have been central to the exercise of U.S. power and the delivery of disaster relief. They have been used to make shows of force, deter adversaries, engage friends and allies, provide humanitarian assistance, and bring airpower to bear against opponents.

A carrier's most potent asset is its air wing. A typical carrier-based air wing today consists of a variety of fixed-wing aircraft (36 F/A-18 Hornets, ten to 12 F-14 Tomcats, six S-3B Vikings, four E-2C Hawkeyes, and four EA-6B Prowlers) and helicopters (four SH-60 and two HH-60 Seahawks). As the F-14 is phased out in coming years, the fighters in the air wing will initially become all F/A-18 and later a mix of F/A-18s and F-35 Joint Strike Fighters. The Navy intends to retain approximately 50 strike aircraft in the carrier air wing as it evolves over time. An extensive network of repair and maintenance, command, control, communications, and intelligence capabilities supports this air wing and the battle group that surrounds the carrier.

In many respects, a carrier is a small city that provides a range of services. Among other things, it makes and delivers freshwater, produces and distributes electrical power, maintains 24-hour-per-day restaurants, operates television stations, provides hospital and dental care, delivers mail, and runs barbershops. These cities are made up of some 5,000 technologically sophisticated men and women who possess a variety of nautical, engineering, aeronautical, electrical, medical, logistical, and warfighting talents.

The military advantages of aircraft carriers are obvious: They can quickly move tactical aircraft and their support to distant theaters of war; respond rapidly with tremendous firepower to changing tactical situations; support several missions at once, with a great number of flights per day; deploy in international waters without having to engage in negotiations with other nations; and remain on assignment for months.

However, as recent events at home and abroad have demonstrated, the nature of conflict is changing, and the United States no longer can consider itself to be an unassailable sanctuary. Moreover, with defense budgets coming under increasing scrutiny, policymakers are under increasing pressure to fully exploit all military assets and to minimize the prospects that assets may be underutilized.

In such an environment, it is likely that aircraft carriers, which are the military's costliest platforms, will be called upon more frequently and be expected to shoulder more duties. With their aircraft, helicopters, and unmanned aerial vehicles; their large open and covered spaces; their significant human resources; and their massive electrical-power-generation capabilities, aircraft carriers represent a significant resource that could be deployed in nontraditional ways.

Historical Nontraditional Uses of Aircraft Carriers

RAND's research teams reviewed past uses of aircraft carriers and projected how and under what circumstances the vessels might be used in the future. For the historical effort, one RAND research group reviewed past employment of the vessels in military operations, concentrating on how they were used in World War II, when the era of today's big flattop carriers came into being, and in subsequent years. RAND's other research group investigated carriers' assignments to past homeland defense missions, to natural disaster response operations, and to other nonhostile endeavors, such as electronic surveillance or spacecraft recoveries.

These historical reviews found that aircraft carriers have been used in a variety of nontraditional combat roles. During World War II, for example, they were used as platforms from which to launch bombers in the Doolittle Raid on Tokyo in 1942, as vessels to transport Royal Air Force and U.S. Army planes to various theaters, and as launch platforms for Army spotter planes throughout the Pacific. In the Vietnam War, carriers were used as electronic intelligence and communications antenna farms. And in later conflicts, they have been used as bases for Army air assault and Special Operations Forces (SOF).

Aircraft carriers also have been used in noncombat roles—as launch platforms for U-2 spy planes, spacecraft-recovery vehicles, troop transports, mobile electric-power plants, and as centers from which to conduct disaster-relief operations.

These historical examples suggest that the Navy has not been shy about using aircraft carriers in alternative ways in the past. These examples also suggest that, as carriers approach the end of their combat service lives, they may be able to accommodate new and different noncombat roles, such as command nodes, communications hubs, or spacecraft-recovery vessels. While such roles may require that older carriers have their catapults removed or go through other modifications, those modifications may extend the carriers' useful service lives by many years.

Uses of Aircraft Carriers in Future Operations

To gauge the nature of the demands that the carrier fleet might encounter in the future, the RAND research groups mapped out 12 combat and noncombat scenarios that they speculated the United States might encounter. The scenarios, which are highlighted below, represent the range of new challenges for which the COGs considered the fleet would need to be prepared. The set of scenarios was not intended to be all-inclusive. Rather, it represents the types of combat and non-combat missions that aircraft carriers could undertake in the future.

Combat Scenarios

- *China-Taiwan crisis*—Set in early 2009, this scenario examined the possibility of the United States coming to the assistance of Taiwan as Taiwan is threatened by the People's Republic of China (PRC).
- *Pakistan coup attempt*—This scenario examined the possibility that a radical group within the Pakistani military attempts to overthrow the government in Islamabad.
- *Korea crisis*—This vignette, set later in this decade, when North Korea might have a dozen or more nuclear weapons, examined some of the issues associated with a confrontation with a nuclear-armed middle-level nation.
- *Crisis in Straits of Hormuz*—This vignette, set late in this decade or early in the next, involved the sponsorship of nonstate terrorist groups by a nuclear-armed Iran.
- *Nigeria civil war noncombatant evacuation*—This vignette examined the capabilities that would be required in a large-scale noncombatant evacuation operation in the wake of a civil war in Nigeria.
- *Colombia insurgency*—This vignette involved the provision of U.S. assistance to Colombia's police and military to counter an insurgency by two major guerrilla groups.
- *Myanmar civil war*—This vignette postulated the provision of U.S. assistance to the Myanmar government pressed by a foreign-backed civil war.

Noncombat Scenarios

- *Nuclear detonation at Long Beach*—This case assumed that a radical nonstate terrorist group has managed to obtain a nuclear weapon, smuggle it into Long Beach, California, aboard a container ship, and detonate it.

- *Atlantic tsunami*—This vignette postulated that a major underwater earthquake occurs in the mid-Atlantic, causing major tidal waves to hit Spain, Portugal, North Africa, and portions of the U.S. East Coast.
- *Volcanic eruption in Hawaii*—This vignette assumed that the volcano of Kilauea on the Big Island erupts with great force, causing massive damage to major portions of the island.
- *San Francisco earthquake*—This vignette assumed that a massive earthquake strikes the San Francisco area with relatively little warning, causing considerable damage to local infrastructure and several thousand deaths and injuries; it also assumed a simultaneous security crisis in Korea.
- *Cuban Mariel-like refugee crisis*—This vignette involved post-Castro civil unrest in Cuba, leading to a massive flood of Florida-bound refugees.

Recommendations Resulting from Scenario Examinations

For each of the scenarios above, RAND examined tasks that the United States might assign to its carrier fleet, assessed the degree to which the fleet's current capabilities could handle them or would need to change, and assessed the operational and technical implications of such changes. RAND identified ten recommendations for the carrier fleet, five related to future combat missions and five to future noncombat missions.

Combat Recommendation: Improve Abilities to Reconfigure Carrier Air Wings

The current air wing is heavily weighted toward strike and anti-air operations. Depending on the situation, carriers will need to alter their mixes of aircraft and, perhaps, bring aboard non-naval aircraft. This concept is not new; non-naval aircraft have operated from U.S. carriers since 1942. Depending on the situation, the normal mix of aircraft might have to be altered, sometimes on short notice or after a carrier has reached its operational area, requiring changes to a carrier's maintenance facilities, weapon storage, and berthing. While such changes would be particularly challenging if the different aircraft came from other services, even additions to the Navy or Marine Corps aircraft complements already onboard would require that a carrier change its mix of spare parts and other key support items.

Combat Recommendation: Increase Carrier Modularity

Today, aircraft carriers can certainly take aboard personnel and aircraft for nontraditional missions, but they are not well suited to act as a base of operations for nontraditional capabilities for extended periods of time. Modularity would enable a carrier to bring aboard new capabilities, in the appropriate mix and in the right quantities, so that it can be an operational base for specific missions. Examples of this modularity concept include containers of spare parts and key maintenance equipment; temporary, modular spaces for use by SOF elements that could deploy aboard ship for extended periods; or modular medical facilities that would increase the ship's organic medical capability.

Combat Recommendation: Obtain Greater Reconnaissance and Surveillance Capabilities

The need for greater long-range, long-endurance, all-weather, stealthy, armed and unarmed intelligence, surveillance, and reconnaissance (ISR) capability came to the forefront in each combat vignette that RAND examined. If carriers were to have the ability to project and sustain, to *at least 500 nautical miles (nmi)*, a persistent ISR capability that includes a mix of sensors (imaging intelligence [IMINT]—electro-optical, radar, and other—and signals intelligence [SIGINT], communications intelligence [COMINT], and electronics intelligence [ELINT]), and the ability to quickly process and disseminate that data, the entire joint force would benefit.

Combat Recommendation: Increase the Ability to Operate at Greater Range and Endurance over Larger Operational Areas

Many of the vignettes (Nigeria, Pakistan, Iran, Myanmar, Colombia) highlighted the fact that aircraft from a carrier, whether manned or unmanned, would have to operate 500 nmi or more from the ship.

This insight is supported by recent operations, such as Operation Enduring Freedom in Afghanistan during 2001–2002, when Navy aircraft ranged far inland on combat and patrol missions. Until Air Force and Marine Corps aircraft could start operating ashore in adequate numbers (a process that required weeks of political negotiations and substantial logistical preparation), aircraft carriers provided the overwhelming majority of tactical aircraft.

Today's carrier air wings would have considerable difficulty maintaining more than a handful of aircraft at distances of 500 nmi or more from the ship. The situation is complicated by the need for persistent coverage in operational areas. Being able to fly a long distance, drop ordnance, and return after spending only a short time in the target area may be appropriate in some situations. In other situations, however, being able to loiter over the area is highly desirable, either for ISR or strike purposes.

Combat Recommendation: Prepare for Operations in a Nuclear Environment

Several cases that we examined—China-Taiwan, Iran, Korea—involved the possible enemy use of nuclear weapons. Such use could include either an overtly lethal and destructive attack by surface or aerial detonation or a high-altitude nuclear detonation to disrupt U.S. command, control, communications, and computer intelligence, surveillance, and reconnaissance (C4ISR) systems. Either development would pose a great challenge to U.S. forces.

Noncombat Recommendation: Enhance Carrier Abilities to Alter the Aircraft Mix Aboard Ship

All noncombat vignettes revealed a need to significantly increase the number of vertical-lift aircraft, a need that stems from the fact that rescue and relief efforts ashore would be in areas where few, if any, airports are available for conventional aircraft. A premium would be placed on helicopters and V-22s, which could bring relief supplies and emergency responders to isolated areas and evacuate badly injured individuals, most of whom would be civilians. At a minimum, naval commanders will need to free up deck and hangar-bay space for more vertical-lift aircraft than the aircraft carrier normally supports.

Noncombat Recommendation: Enhance Carrier Abilities to Provide a Command Center for Key Government Personnel or Agencies

Depending on the level of devastation ashore, some key civilian government personnel may move onto carriers. In the Hawaiian volcano case, for example, the local phone and power systems are so badly disrupted that key officials (the Federal Emergency Management Agency [FEMA] or local government leaders) had to move to the ship for a temporary period.

Noncombat Recommendation: Enhance Carrier Abilities to Provide Medical Facilities for Casualties Brought Back to the Vessel

Most of the vignettes we examined involved massive numbers of civilian casualties. The local medical facilities would almost certainly be overwhelmed, at least initially. In such circumstances, it may be necessary to provide a modular medical capability to enhance the ship's treatment facilities. For example, to augment the normal medical facilities aboard ship, modularized medical containers and additional medical personnel could be flown to the ship and moved into the hangar bay.

Noncombat Recommendation: Improve the Availability of Nonready Carriers

The noncombat vignettes suggest that an aircraft carrier's main value lies in the first few days of a crisis. In this regard, the Navy should consider ways to improve the ability of carriers that recently have returned from deployments or have completed yard periods to deploy on short notice. Although not ready for combat operations, these vessels might be able to put to sea on fairly short notice (one to three days, for example) in order to participate in disaster-relief efforts.

Noncombat Recommendation: Hold Carriers Back from Humanitarian Noncombat Missions When a Major Military Crisis Looms

The San Francisco earthquake vignette provided an example of a major disaster taking place concurrently with a military crisis. Given the limited number of aircraft carriers that would be available for short-notice missions, it is likely that the carrier would be best employed for its primary mission: combat operations. Although the magnitude of the disaster and the availability of other military assets in the threatened region would be key variables, it appears likely that the senior U.S. political and military leadership would want to focus the carriers on combat. The preceding recommendation—improved availability of noncombat-capable carriers to respond to a disaster—would, however, allow the Navy to provide more options to senior civilian leaders who would want as many military capabilities as possible to be available in the event of a major disaster.

These recommendations are offered to the Navy for further consideration. In some cases, there is overlap in the recommendations that pertain to both combat and noncombat operations. For example, the insight that, in the future, carriers should be able to rapidly reconfigure their air wings (and the related recommendation for greater modularity) applies to both combat and noncombat operations. It will, of course, be up to the Navy to decide which recommendations it wants to pursue. This research has shown that nontraditional uses of aircraft carriers are not new: They have been taking place since the 1930s. The insights and suggestions included in this monograph can help the Navy determine how to best employ these powerful and versatile ships.

Acknowledgments

This research could not have been accomplished without the involvement and assistance of many individuals. Brian Persons, Deputy Program Executive Officer (PEO) Carriers, supported and encouraged this work. John Galloway, then–Technical Director, PEO Carriers, offered information, advice, and assistance. Many others brought significant expertise, experience, and insights to our discussions: CAPT (Ret.) J. Talbot "Tal" Manvel, former CVX Program Manager; Owen Cote, Associate Director, Security Studies Program, Massachusetts Institute of Technology (MIT); George Conrad, Alion Science & Technology; many representatives of the U.S. Marine Corps, U.S. Navy, U.S. Coast Guard, U.S. Army Training and Doctrine Command (TRADOC), Special Operations Command (SOCOM), and the Royal Navy; and RAND colleagues John Schank, John Halliday, and Robert Moore.

RAND colleagues James Quinlivan and Robert Murphy offered many constructive comments on earlier drafts, which helped strengthen the final report. We are additionally indebted to Joan Myers for her deft assistance in organizing and formatting the many drafts and to Marian Branch and Nancy DelFavero for their skillful editing.

CHAPTER ONE

Introduction

Defense policymakers have, for a number of years, expressed interest in broadening the roles and reach of aircraft carriers to exploit their capabilities as fully as possible. Because these vessels and their air wings—usually in a Carrier Strike Group (CSG) formation, and sometimes in combination with an Expeditionary Strike Group (ESG)—are some of America's most capable and expensive military assets, civilian and uniformed defense leaders have speculated that opportunities may exist for the United States to leverage the carrier fleet by employing it in new and nontraditional ways. This interest has only heightened since the September 11, 2001, terrorist attacks, as the United States simultaneously has had to adjust to evolving national security responsibilities connected with the Global War on Terrorism (sometimes now referred to as "the long war") and homeland defense and to respond to an array of humanitarian crises, both natural and man-made.

From autumn 2004 until summer 2005, RAND researchers analyzed options available to the U.S. Navy to use aircraft carriers—assigned to either hostile or nonhostile operations—in new and non-traditional roles and missions. On behalf of the Program Executive Office–Aircraft Carriers, Naval Sea Systems Command, RAND explored two fundamental questions: How have aircraft carriers been used in nontraditional ways in the past? What nontraditional roles and missions might aircraft carriers be asked to shoulder in the future?

Relying both on public data and the insights made by government and industry experts in group meetings convened at RAND (which are more fully described later in this chapter), the analysis addressed these questions by cataloging how and under what conditions aircraft carriers have been employed successfully and unsuccessfully in the past and by identifying circumstances that the United States might encounter in the next 20 to 30 years that could require aircraft carriers to be employed out of their traditional role. The analysis also examined alternative ways that carriers could be properly equipped or able to be rapidly equipped with an appropriate mix of capabilities for those roles. The study aimed to help policymakers (1) understand new and emerging military and nonmilitary roles and missions that the aircraft carrier fleet will encounter in the next several decades and (2) identify technical and operational risks and rewards connected with pursuing those new roles and missions.[1]

New or Nontraditional Roles for Aircraft Carriers?

Aircraft carriers and their embarked air wings have been central to the exercise of U.S. naval power since 1942. Time and again, the President has turned to these vessels as the initial policy instrument when the United States has been compelled to project military power or engage in hostile operations. From World War II to today's Global War on Terrorism—playing key roles in four major wars, in operations in Afghanistan and Iraq, and in numerous other hostile and nonhostile missions far and wide—aircraft carriers have been used to make a show of force, deter adversaries, engage friends and allies, provide humanitarian assistance, and bring airpower to bear against opponents.

Modern aircraft carriers, the largest warships ever built, are extremely capable combatants. Each U.S. carrier displaces about 100,000 tons, has a flight-deck area of almost five acres, and is nearly as long as the Empire State Building is tall. Each carrier accommodates more than 5,000 Navy personnel for months at a time. Each is

---

[1]   It should be noted that the study did not specifically include the Navy's large amphibious ships, such as helicopter assault ships (LHAs) and amphibious assault ships (LHDs)—"big-deck amphibs" that are considered specialized amphibious ships for deploying Marine Corps elements. They lack the ability to conduct strike operations with high-performance aircraft. Nevertheless, some of the insights developed in this research could apply to such ships.

expected to operate safely for decades—and, of course, to survive and function as fully as possible in crisis and conflict.

The military advantages of aircraft carriers are obvious: They can quickly move large air forces and their support to distant theaters of war; respond rapidly with tremendous firepower to changing tactical situations; support several missions at once, with a great number of flights per day; and deploy in international waters without having to engage in negotiations with other nations.

However, as recent events at home and abroad have demonstrated, the nature of conflict is changing, and the United States no longer can consider itself to be an unassailable sanctuary. In such an environment, it is likely that aircraft carriers will be called upon more frequently and be expected to shoulder more duties. With their aircraft, helicopters, and unmanned aerial vehicles; their large open and covered spaces; their significant human resources; and their massive electrical-power-generation capabilities, new and existing aircraft carriers represent a significant resource that could be deployed in nontraditional ways. New carriers may also be able to exploit novel capabilities to generate and export electrical power or launch a broader range of air vehicles—capabilities not found in today's Nimitz-class carriers.

Such nontraditional employment would dovetail with today's challenging budget environment and comes at a time when additional capabilities must be provided to existing assets so that they can meet new Navy and Department of Homeland Defense strategies.[2] Aircraft carriers are the military's costliest assets. With defense budgets coming under increasing scrutiny, policymakers are under growing pressure to fully exploit all military assets and to minimize the prospects that assets may be underutilized.

---

[2] Such new strategies include the Navy's still-emerging concepts that will provide a significantly improved ability for joint forces to operate from the sea.

The Use of Concept Options Groups

Recognizing the potential of nontraditional carrier uses, the Navy in 2005 hired RAND to explore possible nontraditional roles for aircraft carriers. Between February and April 2005, RAND created and convened two Concept Options Groups (COGs)—small groups of experienced military and civilian operators and potential users who work together to identify promising ways to employ military might in nontraditional ways—to explore possible nontraditional roles for aircraft carriers. One COG (referred to hereafter as COG-1) explored and identified new ways that aircraft carriers could be used in combat operations; a second COG (referred to as COG-2) examined ways that the vessels could be used in noncombat missions, such as for homeland security or efforts to help the nation recover from terrorist attacks or natural disasters in U.S. territories.

RAND developed and used COGs to good effect in past research projects (Birkler, Neu, and Kent, 1998), in which they were instrumental in helping policymakers explore and identify new and emerging mission needs, technologies, and operational concepts. In this project, each COG was made up of no more than a dozen members, whom we identified with the assistance of the Program Executive Office–Aircraft Carriers, Naval Sea Systems Command. The two COGs operated in parallel, and some of their membership overlapped, depending on the nature of the discussions and the technologies and concepts considered. The membership, which is detailed in Appendix A, included

- experienced military experts from the services and from intelligence elements that might plausibly contribute to the specified mission (COGs 1 and 2)
- broadly knowledgeable technologists drawn from a variety of scientific and engineering backgrounds (COGs 1 and 2)
- senior analysts and planners from RAND and other defense research institutions and from the Department of Defense (COGs 1 and 2)

- federal and state homeland security officials from the U.S. Coast Guard, the Department of Homeland Security, and other federal and state agencies (COG-2).

Each COG convened for three sessions, and each session lasted two days. The sessions were spaced roughly a month apart over three months.

COGs: Reviewing Aircraft Carrier Roles Past and Future

To gain historical perspective, the two COGs reviewed how the United States and other countries have used aircraft carriers traditionally. The combat group (COG-1) focused on past employment of the vessels in military operations, concentrating on how they were used in World War II, when the era of today's big flattop carriers came into being, and in subsequent years. The noncombat group (COG-2) investigated carriers' assignments to past homeland defense missions, to natural-disaster-response operations, and to other nonhostile endeavors, such as electronic surveillance or space-capsule recoveries.

This historical review gave the COGs a good understanding of the types of roles and missions that the carrier fleet has taken on over the past 65 years. While it is likely that some, if not most, of those roles and missions will continue in years to come, it also is likely that new ones will emerge. To gain an understanding of what the carrier fleet might encounter over the next two or three decades, the COGs laid out a dozen scenarios in which aircraft carriers might be expected to play a part. The scenarios, which are more fully described in Chapters Four and Five and in the Appendix, involve the United States in combat and noncombat operations at home and abroad, connected both to military and to homeland defense/humanitarian operations. The scenarios, which take place over the 2008–2020 time frame, are purely speculative, but they were chosen to represent the range of challenges that aircraft carriers might have to overcome. The scenarios are as follows:

Combat Scenarios

- China-Taiwan crisis
- Pakistan coup attempt
- Korean crisis
- Crisis in Straits of Hormuz
- Nigerian civil war noncombatant evacuation
- Colombia insurgency
- Myanmar civil war.

Noncombat Scenarios

- Nuclear detonation at Long Beach
- Atlantic tsunami
- Volcanic eruption in Hawaii
- San Francisco earthquake
- Cuban Mariel-like refugee crisis.

For each scenario, the COGs examined the tasks that the United States might assign its carrier fleet and assessed the degree to which the fleet's current capabilities could handle them. When there was a mismatch, the COGs explored how the capabilities of the carrier fleet would need to change and assessed the operational and technical implications of such changes.

The expert discussions that helped inform the analysis were divided into two major topic areas: (1) nontraditional uses of carriers in combat situations and (2) nontraditional employment of carriers in noncombat missions, such as humanitarian assistance and disaster relief. For each scenario, the participants were provided vignette materials as a read-ahead. The details of the vignettes were reviewed for the group when it convened, then senior RAND analysts served as the group facilitators. The possible roles of aircraft carriers in each of the vignettes were discussed, including what the possible advantages and disadvantages of employing a carrier would be in each situation. The RAND analytic team assembled the insights developed during these sessions and collated them, along with independent assessments made by RAND, into the insights provided in this monograph. The teams included active-duty personnel from the U.S. Navy, the Army, the Marine Corps, Coast Guard, and Air Force. Additionally, civilian analysts participated. Finally, Royal Navy officers from the British Embassy in Washington, D.C., were present for most of the group discussions.

Study Outline

This monograph summarizes the activities, findings, and recommendations of both carrier COGs. Following this Introduction, we devote two chapters to past and current uses of aircraft carriers. Chapter Two reviews the capabilities of aircraft carriers and how the United States has employed them in traditional military operations. Chapter Three describes how the United States used carriers in nontraditional ways in the past. We devote two subsequent chapters to investigations of how carriers might be used in the future. Relying on combat scenarios, Chapter Four investigates how the vessels might be employed in future combat operations; employing a similar scenario methodology, Chapter Five examines how the vessels might be used in the face of noncombat challenges. Chapter Six summarizes our conclusions and recommendations. Lastly, the Appendix details each scenario that the study team used in its analysis.

CHAPTER TWO

Aircraft Carrier Capabilities

To assess how the U.S. Navy might use aircraft carriers in the future, policymakers need to have an understanding of the capabilities of the current fleet of Nimitz-class warships. In the next 10 to 15 years, the degree to which the Navy can take on new combat or noncombat responsibilities will depend, in large measure, on the resources and capabilities that can be provided to the ships that currently are part of the fleet.

This chapter discusses the capabilities that today's U.S. aircraft carriers possess or can call upon as needed. These capabilities fall into several categories. Some capabilities, such as the ability to generate significant amounts of electrical power from a nuclear reactor, are specific to Nimitz-class carriers. Other capabilities derive from the air wing that is connected with a specific carrier or from the surface and subsurface ships that collectively make up a Carrier Strike Group. Still other capabilities, such as satellite communication systems or intelligence-interception systems, can be found elsewhere in the Navy and are shared by many elements of that service.

In combination, these capabilities make the nation's carrier fleet a formidable force today. More than anything else, they provide U.S. policymakers with flexibility. No other asset in the U.S. military arsenal can bring as much freedom of action to U.S. decisionmakers' ability to respond to crises nearly anywhere in the world.

Aircraft carriers and their associated Carrier Strike Groups[1] can operate independently for long periods of time and maneuver in areas to which the U.S. land-based tactical air forces may not have access. This flexibility allowed the United States to overcome access obstacles in operations in Afghanistan in 2001 and in Iraq in 2003.[2] This ability to operate in areas in which an air base is absent or restricted means that carriers can provide varied options to the senior U.S. military and political leadership and to Congress to support U.S. missions, which range from executing humanitarian missions to performing sustained strike operations.

The foundation of a carrier's versatility is the combination of her virtually unlimited range and endurance; her embarked air wing's air-power; her robust communication architecture, which provides for significant command and control capabilities; and the ability to take on mission equipment tailored for the assigned missions. But the carrier offers more: A small city, it provides services ranging from freshwater to an electrical grid, 24-hour restaurants, television stations, hospital and dental care, barbershops, and mail delivery. In addition, a carrier's crew is made up of multitalented, technologically sophisticated men and women who possess a multiplicity of nautical, engineering, aeronautical, electrical, medical, logistical, and warfighting skills. These vast capabilities are why the carrier is the preferred tool in times of crisis for so many decisionmakers. This chapter discusses some of these important features, using a Nimitz-class carrier as the model.

---

[1] A CSG usually comprises an aircraft carrier and its air wing, a cruiser, two destroyers, a fast-attack submarine, and an auxiliary vessel that provides fuel and resupply.

[2] As stated in Larrabee, Gordon, and Wilson, (2004, p. 54), "Turkey's refusal to allow the United States to use its facilities during the Iraq War highlights (the problem of assured access). . . . The U.S. experience in the first few months of operations in Afghanistan in 2001 provides another example of the access problems the United States could face in the future, as it took months to negotiate basing permission from the countries surrounding Afghanistan."

Carrier Air Operations[3]

The most potent asset of an aircraft carrier is its air wing. A carrier is capable of supporting 125 sorties a day, surging up to as many as 200,[4] and can do so for about two weeks before shutting down for one day of maintenance—after which it can do so all over again. The carrier's air traffic control center (CATCC) and primary flight control (PRI-FLY) use the integrated shipboard information system (ISIS), a data management system that collects, distributes, and displays information, to manage flight operations. The carrier crew can launch two aircraft and land one every 37 seconds in daylight, and can launch and land one aircraft per minute at night.

Recent missions in Operations Enduring Freedom and Iraqi Freedom have demonstrated the capability of carrier-based aircraft to support operations over a landlocked theater at distances as great as 750 nautical miles (nmi), with missions lasting between eight and ten hours each. In fact, four Carrier Battle Groups (as they were called in 2001) maintained a sufficient sortie rate to enable a constant presence over the Afghan theater of operations more than 400 nmi away, generating more than 70 percent of all combat sorties in the campaign (Lambeth, 2005). Carrier operations today "provide on-call close-air support, armed reconnaissance and surveillance, airborne command and control, as well as electronic warfare support to the multinational forces in Iraq" and are heavily focused on supporting maritime interdiction operations (MIO) throughout the world, using the air wing's reconnaissance capability to help "detect, disrupt and deter interna-

---

[3]   Unless otherwise noted, primary sources for this section include the U.S. Navy Chief of Information Web site on Aircraft Carriers at http://www.chinfo.navy.mil/navpalib/ships/carriers/; *Jane's All the World's Ships* and *All the World's Aircraft* (available online at globalsecurity.org); and the Naval-Technology Web site at http://www.naval-technology.com/.

[4]   In 1997, the Navy conducted an experiment on USS *Nimitz* designed to maximize strike sorties. During a four-day period, strike sorties totaled 195, 193, 202, and 212. See "Aircraft Carrier Firepower Demonstrated During Exercise" (1997). Although some artificialities were built into the experiment, such as additional personnel and access to spare parts, the sortie generation was a significant accomplishment.

tional terrorist organizations while providing security and stability in the maritime environment [in] the North Arabian Gulf" (Toremans, 2005).

A typical carrier-based air wing consists of 36 F/A-18 Hornets, ten to 12 F-14 Tomcats,[5] six S-3B Vikings, four E-2C Hawkeyes, four EA-6B Prowlers, four SH-60s, and two HH-60 Seahawks. Each provides the following unique capabilities:

- The F/A-18 Hornet is a single-seat, all-weather fighter and attack aircraft capable of supporting strike, counter-air, and close-air support missions and is the workhorse of naval aviation. The newest model, the Super Hornet (F-18 E/F) variant, is currently being placed in the fleet.[6]
- The F-14 D Tomcat is a dual-seat, all-weather fighter aircraft with the ability to support strike operations. It can carry the Sidewinder, Sparrow, and/or AIM-54 Phoenix.[7]

---

[5]   Some carriers have four Hornet squadrons. The Tomcats are being phased out of the Navy's inventory, and the current plan is that they will be out of operation in mid–fiscal year 2007 (FY07).

[6]   Its hard points (tie-down points on the deck) can be loaded with AIM-7 Sparrow, AIM-9 Sidewinder, AIM-120 Advanced Medium-Range Air-to-Air Missiles (AMRAAMs), AGM-84 Harpoon, AGM-88 high-speed anti-radiation missiles (HARMs), AGM-84 Standoff Land Attack Missiles (SLAMs) and SLAM-Expanded Response (SLAM-ER) missiles, and AGM-65 Maverick missiles; Joint Stand-Off Weapons (JSOWs); Joint Direct Attack Munitions (JDAMs); data link pods; Paveway laser-guided bombs; various general-purpose bombs; mines; rockets; and even extra fuel tanks. In a standard interdiction configuration— two SLAM-ERs, two AMRAAMs, two Sidewinders, and three fuel tanks—its range is about 945 nmi. In an air superiority role—operating about 150 nmi from the carrier and equipped with six air-to-air missiles and three fuel tanks—it can remain on station for about 2.25 hours (hr). The F/A-18 is equipped with the Advanced Targeting Forward-Looking Infrared (ATFLIR) or LITENING targeting/laser designation pod for target acquisition and strike, and can also be equipped with the Shared Reconnaissance Pod (SHARP), which provides electro-optical and infrared digital imagery that can be recorded, displayed, and transmitted via data link back to the carrier's intelligence center (Raytheon Technical Services, n.d.).

[7]   In the strike role, the Tomcat can carry MK-83s, MK-84s, laser-guided bombs, and HARMs. Its combat-air-patrol (CAP) loiter time is a little more than 2.05 hr (at a range of

- The S-3B Viking is an all-weather aircraft capable of anti–submarine warfare[8] and anti–surface warfare, electronic support, reconnaissance, and search and rescue. Today, the S-3B most often operates as tanker support for the air wing. The Viking is being phased out of service; it will be replaced by the Super Hornet by 2008.
- The EA-6B Prowler is a long-range, all-weather aircraft with advanced-electronic-countermeasures capability. It provides suppression of enemy air defenses (SEAD) in support of ground or airborne strike operations by jamming enemy radar, electronic data links, and communications. The Prowler can also collect signals intelligence, although its processing capabilities are limited. Recent upgrades, however, have improved the Prowler's ability to geolocate and to link the data to other users. These upgrades are to be in service by the end of 2005.[9]
- The E-2C Hawkeye is an all-weather, airborne early-warning and command and control aircraft. With a crew of five and a sophis-

---

roughly 350 nmi, with two 280-U.S.-gallon drop tanks), and its CAP range (with 1-hr loiter) is about 850 nmi. The Tomcat also can be equipped with a tactical airborne reconnaissance pod system (TARPS), which takes oblique and panoramic film images that can be done on film, so the film must be developed and processed aboard the carrier.

[8]   The S-3 can carry Harpoons, Mavericks, SLAM-ERs, and a wide assortment of conventional bombs and torpedoes. It also has an Inverse/Synthetic Aperture Radar (ISAR), Infrared (IR), and Electronic Support (ESM) system for surveillance and reconnaissance. Its range is about 2,300 nmi, and it can loiter on station (about 150 nmi from the carrier) for about 6 hr. It should be noted that, between 1991 and 1999, the Navy used an S-3 airframe with an expanded intelligence suite, the ES-3A Shadow, as an organic intelligence asset with significant electronic intelligence and communications intelligence-gathering capabilities. Much of the avionics that supported anti–submarine warfare (ASW) has been stripped off the S-3B.

[9]   Prowler can carry the HARM and has shown a capability to counter some improvised explosive devices (IEDs). Its range is about 1,000 nmi, and it can remain on station, unrefueled, for about 4 hr. The EA-6B will be phased out of the fleet beginning in 2009, to be replaced by the EA-18G Growler (a variant of the F-18E/F).

ticated electronics suite, the E-2C coordinates and controls airborne operations ranging from strike operations to search-and-rescue (SAR) missions.[10]

- The Seahawk is a multimission helicopter that can fill anti–submarine warfare (ASW), SAR, drug interdiction, anti–surface warfare (ASUW), lift, and special operations missions. The carrier-based models of SH-60F and HH-60 support ASW/ASUW and SAR missions, respectively.[11]

- The C-2 Greyhound normally operates from a shore facility to support carrier operations; however, it can easily be carrier-based. It can deliver up to 10,000 pounds (lb) of cargo to the carrier and transport personnel and litter patients. The Greyhound has a 1,300-nmi range and is currently undergoing a service life extension program (SLEP) that will keep it in service through about 2020.

Carriers have extensive repair facilities to support both the air wing and the ship, including an aircraft intermediate maintenance department (AIMD), an electronics repair shop, and numerous ship repair shops. AIMD is the backbone of maintenance support for the air wing and can perform advanced repairs on engines, propellers, hydraulics, aircraft structure, avionics, communications, radars, weapons, and other systems to keep the air wing running.

Carriers also have an embarked meteorology department, which develops sophisticated weather forecasts for the air wing and for the ship's navigators.

---

[10] The most recent upgrades to the Hawkeye include a cooperative engagement capability (CEC), which, combined with the shipboard Aegis weapon system, will form the cornerstone of future sea-based theater air and missile defense operations (TAMDO). TAMDOs are designed to work against ballistic and cruise missiles. An advanced version of the Hawkeye, slated to be placed in the fleet beginning in 2011, will have further improvements to its radar and control systems. The maximum range for a Hawkeye is about 1,500 nmi, and on-station time (150 nmi from the carrier) is about 4.5 hr.

[11] Its range is about 380 nmi, and it can carry AGM-114 Hellfire, AGM-119 Penguin, and Mk46 and Mk50 torpedoes.

Command, Control, and Communications, and Intelligence

The aircraft carrier's command, control, and communications systems and its intelligence capabilities are critical enablers in supporting the air wing and entire CSG across its array of missions. More important, because the Navy has yet to come up with a plan to replace its command ships, aircraft carriers will play a much larger role in controlling the battlespace from sea.

Command, Control, and Communications

The carrier's nerve center is the combat direction center (CDC), which controls pictures of the air, surface, undersea, strike, and information warfare, as well as being responsible for protecting the ship with its own self-defense systems.[12] In addition, the embarked flag staff has its own combat center to provide additional command and control guidance in real time.

The backbone of command and control for tactical operations in the Navy is the Joint Tactical Information Distribution System (JTIDS), which is used to provide Link 16 data to the CDC. JTIDS/Link 16 uses a secure, jam-resistant technology to transfer real-time sensor information, identify friend or foe (IFF) information, and geopositional data for aircraft and ships. These data provide situational awareness and battlespace management to the CDC, the ships in the immediate strike group, and other participants in the link, which can include most joint forces. JTIDS operates over line-of-sight ranges up to 500 nmi and can be relayed farther to support additional users in the network.

The Global Command and Control System–Maritime (GCCS-M) is a command-and-control system the Navy uses to provide joint and allied commanders at sea and on shore with an integrated picture. It receives, processes, displays, and maintains geolocation on friendly, hostile, and neutral land, sea, and air forces. It also includes infor-

---

[12] Nimitz-class carriers are fitted with the North Atlantic Treaty Organization (NATO) Sea Sparrow missile, Close-In Weapon Systems (CIWS), and electronic warfare capabilities.

mation provided by JTIDS/Link 16. Additionally, the joint maritime command information system (JMCIS) provides tactical decision aids (TDAs) that the warfighting commander can use in carrying out the operational mission (U.S. Navy, 1998).

Satellite communication suites enable the carrier to access vast information databases worldwide.[13] These systems give the carrier access to the Defense Satellite Communications System (DSCS) for reliable, secure, beyond-line-of-sight information exchange with other fleet units, fixed and mobile joint and allied forces, and Navy command, control, communications, computer, and intelligence (C4I) commands.

Intelligence

Intelligence operations aboard the carrier are designed to support the entire CSG and theater intelligence operations, such as MIO. The embarked staff's intelligence officer (N2), along with the ship's company and air wing intelligence staffs of about 100 personnel, coordinates imagery intelligence (IMINT), signals intelligence (SIGINT), and other national intelligence to provide strike mission planning, indications and warning (I&W), and geopolitical analysis.

The intelligence centers aboard the carrier feature robust systems to support the intelligence mission. The distributed common ground system (DCGS) is a system-of-systems designed to simultaneously task, receive, process, exploit, and disseminate all source intelligence from national, theater, tactical, and multi-intelligence collection assets. The DCGS-Navy[14] (DCGS-N) encompasses not only multiple systems but also the personnel, processes, and training required to operate

---

[13]   Such suites contain permanent extremely high-frequency (EHF), super high-frequency (SHF) (including commercial C-band challenge Athena III and Ku-band satellite), upgraded ultra-high-frequency (UHF), and Global Broadcast System (GBS).

[14]   DCGS subsumed the capabilities of the joint fires network (JFN).

the systems and provide intelligence analysis in support of operational decisionmaking (U.S. Navy, 2003; "DCGS-N Budget Item Justification Sheet," n.d.).

In addition to DCGS and its supporting subsystems, intelligence centers aboard the carrier use the integrated broadcast service (IBS) and the joint tactical terminal (JTT), which integrate various SIGINT data to support I&W, surveillance, targeting data, and SAR requirements of operational commanders and targeting staffs across all warfare areas. IBS and JTT are capable of sending data via other communications paths, such as SHF and EHF (U.S. Navy, 1998).

It should be noted that the carrier's intelligence team is configured to support the air wing, ship, staff, and strike group. Despite the broad capabilities outlined above, the intelligence team has limited ability and expertise to support ground operations or special operations. The intelligence team's ability to support these missions improves when the carrier carries a Marine squadron of F/18s or when personnel with Naval Special Warfare experience are part of the crew complement. Additionally, the intelligence team has a wide variety of shore resources to which it can turn for greater depth in these areas.

Other Aircraft Carrier Capabilities: Toward the Nontraditional

A carrier, then, is a floating city with a broad array of services. Therefore, a Nimitz-class nuclear propulsion plant not only can propel a 100,000-ton ship through the water at speeds in excess of 30 knots (kt), it can also power electrical generators supporting both a mobile airport that uses complex, high-powered electronic equipment and a complement of ship and air-wing crews totaling more than 5,000 individuals. The plant also provides steam for catapults and for desalinization units capable of producing more than 400,000 gallons of freshwater per day. With more than 20 years' worth of fuel at normal operating tempo, the ship's tactical range is essentially unlimited. Nimitz-class carriers normally carry enough supplies to remain at sea for 90 days without resupply, with the exception of combat consumables (i.e., aviation fuel and ammunition).

There is enough space on the carrier to support a crew of 3,200, an air wing of nearly 2,500 personnel, and an admiral's staff of more than 100 personnel. A carrier's food services department produces 18,000 to 20,000 meals per day. It has a well-equipped, 50-bed hospital manned by six doctors, at least one of whom is a surgeon. There also is a dental clinic aboard, with five dental officers who normally see as many as 70 patients per day. The capabilities of a Nimitz-class carrier are at least twice those of the other elements of U.S. strike groups, as Table 2.1 sets forth.

All of these capabilities are designed to support the crew, but they have also been employed historically to provide relief in all types of humanitarian and national-security crises, as the next chapter illustrates.

Table 2.1
Comparison Among Nimitz-Class Carriers and Other "Flatdecks" in the U.S. Navy

Capability	Nimitz-Class CVN	Tarawa-Class LHA	Wasp-Class LHD
Displacement (ton)	97,000	39,400	40,650
Length (ft)	1,092	820	844
Maximum speed (kt)	Over 30	24	23
Crew	5,900	2,900	3,000
Range (nmi)	Tactically unlimited	10,000	9,500
Maximum aircraft	90	35	35
Air operations	24 hours per day, sustainable for approximately 2 weeks	12 hours per day, sustainable for about 3 days	12 hours per day, sustainable for about 3 days
Freshwater production (gallons per day)	400,000	140,000	200,000
Hospital beds	50	60	60

NOTE: CVN = nuclear aircraft carrier.

CHAPTER THREE

Historical Nontraditional Uses of Aircraft Carriers

The Navy has used aircraft carriers for what could be considered nontraditional uses—the employment of the ship for missions other than to directly support naval operations, such as strike, escort, and defensive protection of other ships or forces ashore—ever since the early 1930s. This chapter reviews a number of the past uses of aircraft carriers to provide a historical context for possible future missions.

Nontraditional Combat Employment of Aircraft Carriers

The Doolittle Raid—April 1942[1]

The Doolittle Raid is one of the earliest, and most famous, nontraditional missions that a U.S. aircraft carrier conducted. USS *Hornet* (CV-8) loaded 16 U.S. Army B-25 medium bombers in port on the West Coast (by crane; the Army bombers could not fly aboard the ship) and, under conditions of great secrecy, carried them within striking range of Japan. Because it could not launch or recover any of its normal aircraft with the Army bombers on its deck, the *Hornet* was escorted by USS *Enterprise* (CV-6) and various cruisers and destroyers. However, Japanese fishing boats, deployed in a picket line several hundred miles east of the home islands, detected the approaching U.S. force. Before *Hornet*'s escorts sank them, the picket vessels sent radio messages back to Tokyo, alerting the Japanese military of the presence of the

---

[1] The name "Doolittle Raid" comes from the name of the Army mission commander, Lieutenant Colonel James Doolittle.

Americans. That event forced the task force to launch the Army bombers hours before they were scheduled to take off. Despite the warning provided by the fishing vessels, the Japanese were still taken by surprise when the Army planes arrived over Tokyo and several other cities in daylight. The raid resulted in minuscule damage to Japanese facilities. No Army planes were shot down by the Japanese defenses, but all had to crash-land (mostly in eastern China) as a result of fuel shortage—an effect of the premature takeoff.

Although the physical damage to Japan was minimal, the Doolittle Raid was a tremendous morale boost to the U.S. public. The previous four months had seen one defeat after another, including the bombing of Pearl Harbor and the Japanese overrunning of much of Southeast Asia, including the Philippines, then a U.S. possession. In addition to lifting the spirits of the U.S. military and public, the Doolittle Raid so humiliated the Imperial Japanese Navy that the Japanese quickly finalized their plans to draw out the U.S. Pacific Fleet through an attack on Midway Island in early June 1942. History records that battle as the decisive turning point of the Pacific War, where four Japanese aircraft carriers were sunk—in large part the effect of the Doolittle Raid of goading the Imperial Navy into an ill-considered offensive (Morison, 1968b, pp. 387–398).

Saving Malta—April/May 1942

An important strategic location of World War II, astride the German-Italian supply lines running from Italy to North African ports, was the British-controlled island of Malta, located in the center of the Mediterranean Sea, roughly 100 miles south of Sicily. British aircraft, submarines, and surface ships based on Malta inflicted considerable loss on Axis transports moving men and supplies from Italy to their forces fighting the British in Libya and Egypt. By spring 1942, Axis leaders had suffered enough from Malta and made plans to seize the island by airborne and amphibious assault.

Spring 1942 was a desperate time for the United States and Great Britain. With the Japanese offensive in the Pacific, a crisis situation in the battle against the German U-boats in the Atlantic, and major battles raging in North Africa, Allied resources were stretched thin across the globe. Nevertheless, Malta had to be supported, and that meant getting more fighter planes and other supplies onto the island, which was being heavily bombed by German and Italian aircraft in preparation for the planned invasion. With no British carriers available at that time, the U.S. Navy was asked to support the mission to rearm Malta with Spitfire fighters.

Twice, on April 20 and May 7, USS *Wasp* (CV-7) steamed within range of the island. The speed of the carrier allowed it to rapidly return to the British Isles after the first mission, pick up the second load of fighters, and return to the Mediterranean in less than three weeks. A total of 105 Royal Air Force (RAF) Spitfires took off from the carrier and landed on the island. Both groups of aircraft had been loaded aboard the ship by crane in Scotland. Note that the aircraft were not American, nor were they from Britain's Royal Navy. The pilots were RAF personnel, new to carrier operations.

The additional Spitfires proved decisive to the defense of Malta. The Germans and Italians could not establish air superiority over the island. This reality, in addition to the pressing need to send more troops directly to North Africa, caused the invasion to be cancelled. Within a few months, Malta had regained enough offensive strength to allow it to resume its vital role of interdicting Axis supply routes to North Africa (Morison, 1970, pp. 193–197).

Operation Torch—November 1942

Troop landings in Vichy French–controlled Morocco and Algeria were the first Anglo-American offensive operations in the European/ Mediterranean area in World War II. With the Germans and British locked in combat in Egypt, the Torch landings were intended to seize the western portion of North Africa, thus taking the German-Italian forces from the rear. There was only one British-controlled airbase in the western Mediterranean in late 1942, the outpost at Gibraltar, which had limited capacity and was needed to support the attack on Algeria. Therefore, all the initial air support for the assault on Morocco had to be carrier-based. The plan included provisions for the rapid seizure of French airfields along Morocco's coast, thus allowing land-based aircraft to use those bases. To bring short-range Army fighters close to the Moroccan coast, Navy carriers had to be used as transports.

When Task Force 34 departed from Norfolk, Virginia, on October 23, 1942, the force included USS *Chenango* (CVE-28). Aboard the 24,000-ton escort carrier were 76 U.S. Army P-40 fighters. One of the best French airfields was located close to the coast, near Port Lyautey, north of the capital of Rabat, Morocco. Army assault troops, supported by Navy aircraft from other carriers and by gunfire from ships, went ashore early on November 8. By November 10, the airfield was in U.S. hands and the Army fighters began flying off the *Chenango* to land at the base. Thus, land-based aircraft were available to supplement the Navy's fighters, dive-bombers, and torpedo planes aboard the other carriers that were still operating offshore. This innovative use of an aircraft carrier allowed Army aircraft to quickly start using bases ashore, instead of having to be slowly ferried in from the limited-capacity base at Gibraltar that was, as mentioned, fully committed to supporting simultaneous operations in Algeria (Morison, 1968a, pp. 37, 118–119, 131).

Transporting Army and Marine Corps Aircraft—1942/1945

As the United States started its drive to eject the Japanese from the areas they had seized in the first year of World War II, aircraft carriers were critical to U.S. success. The distances in the Pacific were so great that most World War II–era aircraft, especially relatively short-range fighters and light bombers, often could not cover the distances from one island group to another.

Once a base had been won from the enemy, U.S. commanders wanted to quickly bring land-based aircraft forward to protect the base and, range permitting, start to strike the next set of Japanese-held locations. When the distance was too great to allow Army or Marine Corps aircraft to fly to the just-seized base, Navy carriers were used to bring the short-range aircraft forward. This mission was usually performed by slow, but numerous, escort carriers. During the course of the war, roughly 90 escort carriers were produced. Most were built on tanker or oiler hulls, and they displaced from 12,000 to 24,000 tons or more, depending on the class. When used for tactical missions, escort carriers typically carried 18 to 27 Navy aircraft. When performing transport duties, however, they would carry a much larger number of aircraft, often two to three times the ship's normal complement (as, for example, USS *Chenango* had done during Operation Torch, when it carried 76 Army fighters). Aircraft were parked closely together on the flight and hangar decks, with enough space left on the forward part of the flight deck to permit the first aircraft an adequate takeoff run.

The use of carriers to transport non-Navy aircraft to distant bases had actually started before the Japanese attack on Pearl Harbor. Indeed, USS *Enterprise* (CV-6), destined to be the most decorated Navy ship in World War II, was fortunate to be on a mission to transport Marine Corps fighter planes to Wake Island the week before the Japanese attack. Had *Enterprise* not been conducting this operation, she would have been at Pearl Harbor when the Japanese attack took place and would almost certainly have been sunk.

The use of the escort carriers for this purpose freed the fast fleet carriers for operations against the main Japanese bases and for actions against the enemy fleet. During the course of the war, Navy escort carriers transported thousands of Army and Marine Corps aircraft, mostly in the large Pacific theater.

Army Spotter Planes Aboard Ship—October 1944

The U.S. invasion of the Philippines in 1944 was one of the largest naval operations in history, assembled of a naval force of nearly 850 ships of all types to transport and protect a multi-corps Army invasion force that landed on the island of Leyte in the central Philippines. Among the ships were 16 escort carriers whose air complements had the missions of flying anti-submarine patrols and providing air support to the Army forces ashore until Army aircraft could start operations from land bases. To help coordinate with Army units fighting on Leyte, some escort carriers (grouped in three small task forces and positioned to the east of the island) carried a number of U.S. Army artillery spotter aircraft that initially flew from the ships and later deployed ashore.

Communications, Electronic Intelligence, and Command Platform— Vietnam, 1960s

When U.S. participation in the Vietnam War got under way in the early 1960s, the Navy still possessed a large number of World War II–era ships. Some of these vessels, including some of the remaining escort carriers that had been produced in such large numbers during World War II, were converted for other uses.

For example, USS *Annapolis* (AGMR-1, a communications ship), formerly USS *Gilbert Islands,* CVE-107) and USS *Arlington* (AGMR-2, formerly USS *Saipan,* CVE-48) became major communications relay platforms for U.S. forces in Vietnam. The ships were provided with a large array of antennas and other communications equipment, taking advantage of the flight-deck space.

These ships were used off the coast of Vietnam to support 7th Fleet operations. They provided message handling and relay, and assisted other ships in repairing and better utilizing their communications equipment. Later, in 1968, the *Arlington* supported the spacecraft-recovery program, discussed later in this chapter, serving as a communications relay ship.

Base for Army Air Assault Forces—1994

In September 1994, USS *America (*CV-66) and USS *Eisenhower* (CVN-69) had the unusual mission of transporting Army forces to Haiti in support of Operation Uphold Democracy. Aboard the *America* were elements of the Joint Special Operations Command and helicopters of the 160th Army Special Aviation Regiment. Meanwhile, the *Eisenhower* carried most of the 1st Brigade Combat Team of the Army's 10th Mountain Division from Fort Drum, New York. Aboard *Eisenhower* were numerous Army helicopters, to be used for an opposed air assault operation into Haiti.

On September 19, elements of the 10th Mountain Division conducted an air movement into Haiti from the *Eisenhower*. Fortunately, last-minute political negotiations led to the peaceful stepping down from power of Haiti's military leaders. *Eisenhower* briefly remained in the area to support the movement of Army forces into Haiti, then returned to normal operations in October.

Until mid-October 1994, the *America*, with more than 2,000 Army, Air Force, and Marine Corps personnel aboard, including their helicopters, supported operations on Haiti, launching more than 400 sorties from its decks.

Operation Uphold Democracy was an example of the concept of *adaptive joint force packaging*. As articulated by then–Atlantic Command Commander Admiral Paul David Miller, this concept called for the flexible and even nontraditional use of the assets of all the services to accomplish a mission. The absence of an air threat in Haiti, combined with the need to move considerable numbers of Army and Marine forces to the island, provided an excellent opportunity to use carriers in a nontraditional manner to accomplish the mission.

Platform for Special Operations Forces

Navy aircraft carriers have been used as platforms for Special Operations Forces on a number of occasions. In one of the most publicized examples, Operation Eagle Claw, the April 1980 attempt to rescue the U.S. hostages being held by radical students in Tehran, Iran, eight Navy RH-53D helicopters, manned by Marine flight crews, launched from USS *Nimitz* (CVN-68) in the north Arabian Sea as part of the rescue attempt. Although this operation was unsuccessful, the carrier proved to be an ideal launch point for the helicopters, given the secrecy of the operation and the sensitive political situation in the region.

Other examples followed. Most recently, USS *Kitty Hawk* deployed from its base in Japan to support Special Operations Forces operating in Afghanistan in late 2001 and early 2002. While other carriers operating in the north Arabian Sea provided traditional air capabilities, the *Kitty Hawk's* primary mission was to serve as a platform and base for Special Operations elements that flew northward to Afghanistan.

Nontraditional Uses of Aircraft Carriers for Noncombat Operations

In addition to the nontraditional combat-related uses of aircraft carriers described above, there are also examples of aircraft carriers being used in nontraditional ways to support noncombat operations. As with the combat-related examples, this list is not all-inclusive; rather, it is intended to provide a sample of the range of mission types for which carriers have been used.

Platform for U-2 Spy Planes—1960s

In 1963, the Central Intelligence Agency (CIA) established Project Whale Tale, whose goal was to base and launch U-2 high-altitude reconnaissance planes aboard aircraft carriers. The main advantage of basing U-2s aboard ship was that the United States could then avoid, at least in some circumstances, having to ask other nations for basing rights in order to conduct missions. The first U-2 flight was from USS *Kitty Hawk* (CV-63) in August 1963. A U-2C that had been loaded aboard the ship in San Diego launched without catapult assist, carrying a full load of fuel, in just over 320 feet of deck space. However, the subsequent landing attempt was not successful. The experience led the Navy to modify three other U-2s, adding stronger landing gear, arresting hooks, and wing spoilers. These aircraft became known as U-2Gs.

In March 1964, the first successful U-2 landing on a carrier took place on USS *Ranger* (CV-61). Two months later, a U-2G performed an operational mission from the *Ranger*. During this period, the modified U-2 was used to monitor French nuclear tests that were being conducted in the South Pacific.

In 1967–1969, another U-2 variant, the U-2R, operated from USS *America* (CV-66). This version of the aircraft had twice the range and four times the payload of the earlier models. The U-2R had folding wing tips to make it more compatible with carrier operations. This version of the aircraft was able to move on the ship's elevators and enter the hangar bay (see Pendlow and Welzenbach, 1988, pp. 247–249).

Powering a City—Tacoma, Washington, 1930

In the early 1930s, the Navy possessed only three aircraft carriers: the converted coal collier *Langley* (CV-1); and two converted battle cruisers, *Lexington* (CV-2) and *Saratoga* (CV-3). Originally designed as fast battle cruisers, *Lexington* and *Saratoga* had very powerful propulsion plants, capable of producing roughly 210,000 shaft horsepower, enough power to give the two large (40,000-ton-plus) ships a top speed of nearly 34 kt. For many years, the two carriers were the fastest large ships in the Navy.

The amount of power available in these ships resulted in a very early, innovative use. In late 1929, the city of Tacoma's power system failed. The Tacoma area experienced a drought, which diminished water in the nearby dams that were hydroelectric sources of power to the city. Local businesses had to start laying off employees, and the nearby Army base at Fort Lewis also felt the effect of the dramatic reduction in power output from commercial sources. Tacoma appealed to President Herbert Hoover for help, and the matter was passed to the Navy. After initially turning down the request, the Navy ordered the *Lexington* to Tacoma. For one month, the ship provided roughly 30 percent of the city's electrical power. See Figure 3.1.

Troop Transport at the End of World War II

At the end of the Second World War, the United States had literally millions of military personnel deployed around the globe. While some were retained within operational areas to perform postwar-occupation duties, most were transported back to the United States to be demobilized and released from military service. Since few airplanes of the era could fly transoceanic distances, and the personnel to be redeployed were vast, ships became the primary means of transporting military personnel back home. Aircraft carriers were included in the fleet of ships used for this purpose.

For example, the veteran USS *Enterprise* was available for this mission. Beginning in November 1945 and proceeding into 1946, the ship made several voyages back and forth to Europe. More than 10,000 U.S. military personnel, mostly Army, returned to the United States aboard *Enterprise* in what were called the "Magic Carpet Ride" voyages. In late 1945, USS *Independence* (CVL-22, a light fleet carrier) joined the "Magic Carpet" fleet (see U.S. Navy, http://www.chinfo.navy.mil/navpalib/ships/carriers/histories/cv06-enterprise/cv06-enterprise.html).

Figure 3.1
USS *Lexington* Supplying Power to Tacoma, Washington, 1930

SOURCE: Photo courtesy of Tacoma Public Utilities (http://www.
washington.historylink.org/essays/output.cfm?file_id=5113).
RAND MG448-3.1

Spacecraft-Recovery Vessels—1960s and 1970s

When the United States' manned space program started in the early
1960s, the Navy immediately became involved in the critical task of
recovering returning spacecraft and their crews. The technology of
the era did not permit returning space capsules to come to earth on
land; water landings were necessary. For such landings, the Navy used
a number of older, World War II–era Essex-class aircraft carriers as
recovery vessels.

In 1961, USS *Randolph* (CV-15) served as the recovery ship for the
second U.S. Mercury space mission, recovering pilot Virgil Grissom
upon his return from a suborbital flight. In February of the next year,
the *Randolph* recovered John Glenn, the first American to orbit the
planet in space, following his mission.

As the U.S. space program developed during the early 1960s, the Navy retained the mission of recovering the returning manned spacecraft. Since aircraft carriers were fast and could land and operate considerable numbers of helicopters, they were the platform of choice for these operations. For example, in 1969, USS *Hornet* (CV-12, the second carrier with that name; the first *Hornet* had been sunk in action against the Japanese in October 1942) was the recovery vessel for the first and second Apollo moon missions, *Apollo 11* and *12* (see U.S. Navy, http://www.chinfo.navy.mil/navpalib/ships/carriers/histories/cv15-randolph/cv15-randolph.html).

Figure 3.2 shows a recovery effort that took place three years later, when USS *Ticonderoga* met with *Apollo 17* after it splashed down in the Pacific in 1972.

Disaster-Relief Operations

Navy ships have long played an important role in disaster-relief missions. Carriers were used for this purpose on a number of occasions, both in support of domestic U.S. needs and to help foreign disaster victims.

For example, in 1954, USS *Saipan* (CVE-48) supported relief efforts in the Caribbean following hurricanes that struck the island of Hispaniola. Food, water, medical and other supplies, and personnel were all brought to the devastated area by the ship. In 1955, the same ship provided assistance to Mexico following flooding in the Tampico area.

The Navy conducts these types of relief operations to this day. Its most recent operation took place in the months following the massive underwater earthquake that struck southwest of the island of Sumatra in Indonesia in December 2004. Waves up to 60 feet high struck the southern Sumatran coast, and smaller, but still powerful, waves hit the shores of Thailand, India, and Sri Lanka. The death toll, estimated at roughly 280,000, made this one of the worst natural disasters in modern times.

The U.S. Navy responded by dispatching USS *Abraham Lincoln* (CVN-72) and other ships to the area. For weeks, the *Lincoln* provided

Figure 3.2
USS *Ticonderoga* Recovering *Apollo 17*, 1972

SOURCE: Photo courtesy of National Aviation and Space Administration (NASA)
(http://www.hq.nasa.gov/office/pao/History/alsj/a17/ap17-S72-55974HR.jpg).
RAND MG448-3.2

life-saving relief to devastated coastal communities. The ship's aircraft
complement was altered, with more helicopters being flown aboard in
lieu of fixed-wing aircraft. Food, medicine, and relief personnel were
all flown ashore by the *Lincoln's* helicopters. See Figure 3.3.

Figure 3.3
USS *Abraham Lincoln* Crew Members Readying Tsunami Relief Supplies, 2005

SOURCE: U.S. Navy photo, taken by Photographer's Mate 2nd Class Seth C. Peterson (http://www.news.navy.mil/management/photodb/photos/050109-N-0057P-014.jpg).
RAND *MG448-3.3*

Lessons from Past Nontraditional Uses of Carriers

The examples of nontraditional uses of aircraft carriers cited in this chapter provide an overview of how carriers have been used in the past and demonstrate some still-appropriate possible alternative future uses of these ships.

It is important to remind readers that, during World War II and throughout the Cold War, aircraft carriers were primarily strike and air superiority platforms. Although used occasionally in nontraditional roles, the carrier has normally operated as a conventional combat system that has an aircraft mix optimized for strike and offensive/defensive counter-air missions. This is still true of today's carriers.

It is noteworthy that carriers have been used to transport non-Navy aircraft for many years. Indeed, aircraft carriers performed that mission frequently during World War II, when the range of most land-based aircraft was limited. Today, on the one hand, the range of most fixed-wing aircraft (especially when combined with aerial refueling) is far greater, thus reducing the need for carriers to perform transport or ferry missions. On the other hand, most Army, Air Force, and Marine Corps rotary-wing aircraft still have relatively short ranges. The 1994 intervention in Haiti, in which Army helicopters were carried aboard Navy carriers (and were prepared to conduct assault operations if required), shows how the technique first used in World War II remains valuable.

During World War II, when the Navy comprised thousands of ships, including scores of aircraft carriers, small escort carriers could be used for most nontraditional missions. Today, with only 11 carriers in the fleet, alternative uses would have to be performed by the same ships that focus mostly on traditional naval missions, such as sea control, fleet defense against standoff bomber attack, and strike operations. Nevertheless, several recent operations—such as the Haiti example or the use of the *Abraham Lincoln* off Sumatra in late 2004/early 2005—show that if the conditions are right, the air wing of a carrier can be modified to allow more helicopters aboard to perform different missions.

History suggests that very different roles, including the use of carriers as command nodes or communications hubs or as spacecraft-recovery vessels, may be available as the carriers approach the end of their service lives. Some of these roles would require that older carriers be modified in some way (such as removing the catapults) to tailor them to these other missions. Additional years of useful service might be obtained through such methods.

The historical examples provided here are intended to provide a context and useful data points for how carrier ships have been used for nontraditional missions. They serve to show the flexibility of this type of warship and will help guide us in our examination of possible future missions.

CHAPTER FOUR

Uses of Aircraft Carriers in Future Combat Operations

This chapter examines potential uses of aircraft carriers in future combat situations. It uses as the basis of its analysis a number of scenarios, or vignettes, of a wide variety of operations in which U.S. decisionmakers may choose to use aircraft carriers. RAND prepared these vignettes to include both "high-end" crisis situations against powerful opponents and lower-intensity missions involving U.S. forces, such as noncombatant evacuations and counterinsurgency operations. Importantly, several of the cases include the possible use of nuclear weapons, either by terrorists on a relatively small scale or by an enemy in a more strategic context.

Including nuclear-weapons scenarios was important because of a strategically dangerous trend: the emergence of "middle-level" powers with nuclear weapons. In this case, *middle-level* refers to nations that are far less powerful than such peer-level competitors as the former Soviet Union, but that are still major regional powers with considerable military capability. North Korea, Pakistan, and India all now admit to having a nuclear arsenal. It would be prudent to assume that Iran will gain a nuclear-weapons capability in the not-too-distant future. This changing situation means that the U.S. military must plan for the possibility of engaging in combat with nations that have a modest nuclear capability.

This chapter is organized in two parts. The first part provides a short overview of each of the combat vignettes (the Appendix details all of these vignettes). The second part presents the insights that RAND derived from the Concept Options Group's examination of the vignettes.

Overview of Combat Vignettes

China-Taiwan Crisis

Set in early 2009, this vignette examined the possibility of the United States coming to the assistance of Taiwan when the People's Republic of China (PRC) threatens it. The U.S. response is overwhelmingly air and naval, and aircraft carriers are a key component of the U.S. capabilities rushed to the area. This situation is truly the high-end or worst-case future scenario for the United States, since the Chinese military has far more power than nearly any other potential opponent. The vignette postulates that the PRC has a reasonable amount of long-range reconnaissance and surveillance (including satellites) capability that could allow it to locate maneuvering U.S. naval forces. Additionally, the PRC has a large number of surface ships, submarines, and strike aircraft that could be used to attack U.S. naval units. Its land-based air defense capability is also formidable.

The vignette includes the possibility of the PRC employing high-altitude nuclear detonations (HANDs) to disrupt U.S. command, control, communications, computers, intelligence, surveillance, and reconnaissance (C4ISR) systems. For example, it postulates that the PRC could use a HAND burst to disrupt the electronics of the ships in a Carrier Strike Group and then follow up the detonation with a large air strike to attack the ships while they are still recovering from the effects of such an attack.

Another aspect of this vignette involves the possibility that the PRC could threaten Japan, demanding that Japan prohibit any U.S. use of its bases.

Pakistani Coup Attempt

This vignette examined the possibility that a radical group within the Pakistani military attempts to overthrow the government in Islamabad.

Although the coup attempt fails, the rebels seize one or more nuclear-weapons storage sites and a number of missile launchers. The Pakistani government asks the United States for assistance in the form of intelligence, surveillance, and reconnaissance (ISR), precision strike, and Special Operations Forces liaison personnel to assist in its attempts to quickly retake the storage facilities and prevent the launch or removal of nuclear weapons. Strike and reconnaissance aircraft or unmanned aerial vehicles (UAVs) from carriers operating in the Indian Ocean are a key U.S. capability that can assist the Pakistanis.

The vignette highlights the need for the United States to quickly establish liaison with both Pakistani and Indian authorities. In this situation, U.S. forces would provide detailed, real-time, persistent, all-weather ISR support to Pakistani forces, as well as precision-strike assets that the Pakistani military would lack. It should be pointed out that support by current and projected long-endurance UAVs or manned ISR aircraft cannot be provided unless those systems operate below any cloud layers, which thus makes them subject to attack by man-portable air defense systems (MANPADS) and other air defenses.

Korean Crisis

This vignette, set later in this decade, when North Korea might have a dozen or more nuclear weapons, examined some of the issues associated with a confrontation with a nuclear-armed middle-level nation. In this vignette, the North Koreans are being heavily pressured by nations in the region to give up their nuclear capability. Rather than submitting to this economic and diplomatic pressure, the North attempts to use its small nuclear capability in a coercive manner to overturn the sanctions that are in force against them. An important element of this vignette is the possibility that public opinion in South Korea might be strongly against provoking the North. Therefore, when the situation reaches the crisis point, public pressure might force the Seoul government to deny the United States the right to conduct offensive missions against the North, at least until the South has been directly attacked by the North. Similar to the China-Taiwan case, Japan would be threatened by the possibility of North Korean nuclear strikes should U.S. forces be permitted to use Japanese bases for strikes against the North.

With the possibility of South Korean, and even Japanese, bases being closed to U.S. use, the role of aircraft carriers would be critical. The Navy's surface combatants could be given important missile defense missions to help protect both South Korea and Japan. Such missions could include the need for boost-phase intercept of North Korean missiles as they are launched. Therefore, Navy surface combatants would have to be positioned off the North Korean coast, thus making them vulnerable to attack from the North's aircraft and submarines. The protection of surface combatants performing missile defense, as well as traditional strike operations against targets ashore, would be an important mission for aircraft carriers. An alternative is to have the carrier stand off and operate high-altitude, long-endurance (HALE) aircraft equipped with air-to-air missiles adapted for boost-phase intercept.

The Straits of Hormuz

This vignette, set later in this decade or early in the next, is another situation in which the United States might have to confront a nuclear-armed regional opponent. Here, however, the Iranians are assumed to be sponsoring nonstate terrorist groups in an attempt to undermine the U.S. position in the region. One of the terrorist groups conducts a deadly attack against a hotel in the region at which a large number of Westerners are staying. Hundreds are killed, including a large number of Americans. Shortly thereafter, various intelligence sources confirm that the terrorists were Iran-sponsored and Iran-armed. This information leads the United States to consider air strikes against various key Iranian facilities, including the command centers of Iranian agencies known to be supporting the terrorists and portions of Iran's nuclear infrastructure.

This scenario was set to take place at a time when Iran likely possesses several batteries of modern, high-quality surface-to-air missiles (for example, SA-10 and SA-15), which are assigned to defend key political, economic, and military installations. Importantly, the Iranians also have the ability to strike many nations in the region with missiles, some of which might be nuclear-armed. Additionally, the Iranians possess a large number of small surface craft and coastal defense antiship weapons, as well as "smart" sea mines, which they can use to close the Straits of Hormuz. Indeed, in this vignette, the Iranians threaten to take such a step in the event of a U.S. strike against them.

Unlike the Korean vignette, in which U.S. naval forces would have considerable freedom to maneuver, the U.S. Navy would have to decide whether to keep surface combatants—and aircraft carriers— inside the Persian Gulf in this scenario, given the likelihood that Iran would quickly move to close the Straits. Additionally, the threat of Iranian missile attack may force some of the nations in the region that normally are friendly to the United States to deny basing rights to U.S. forces. The possibility of reduced ashore basing increases the importance of aircraft carriers, but the geographic constraints in this region pose challenges not found in the Pacific vignettes.

Nigerian Noncombatant Evacuation

The RAND team designed this vignette to examine the capabilities that would be required in a large-scale noncombatant evacuation operation (NEO). It postulates that a civil war has broken out in Nigeria, with the mostly Christian south pitted against the primarily Muslim north. Elements of the Nigerian armed forces have gone over to the Muslim rebels, and there are undisciplined militia groups on both sides. Several thousand foreign civilians reside in the country, and most of them are associated with oil-industry facilities near the southern coast. However, there are also pockets of U.S. and other foreign nationals much deeper inland. The distance from the coast to the northern regions exceeds 500 miles. It is likely that various military and paramilitary groups in the country would oppose the presence of foreign troops in the country. Indeed, there have been instances of violence against Americans and other foreign nationals who are attempting to flee the country.

In addition to U.S. military units moving to the area to conduct the evacuation, other nations, primarily the United Kingdom and France, have dispatched air and naval forces to the region. Although negotiations are under way to gain basing rights in nearby nations, approval will take some time. Until these neighboring nations express a willingness to allow aircraft from the United States and other states to use their bases and airspace, the evacuation has to be conducted by naval forces and long-range aircraft.

Colombian Insurgency

Set later in this decade, the Colombian vignette is, like the Nigerian one, a low-intensity situation. In this vignette, Colombia's two major leftist guerrilla groups, the Revolutionary Armed Forces of Columbia (FARC) and the National Liberation Army (ELN), have obtained a variety of advanced infantry weapons and are starting to prevail against the country's police and military forces. The Colombian government sends an urgent appeal for help to the United States, requesting direct military assistance to help avert a collapse. The United States elects to send additional military supplies and equipment, Special Operations Forces (SOF) advisers, and air support. Owing to political sensitivities in the United States and among Colombia's neighbors, the United States must minimize its presence in Colombia. Therefore, carrier-based aviation will be the primary means of providing reconnaissance, surveillance, and strike missions to support the Colombian forces.

This vignette illuminates issues associated with a carrier operating in a low-intensity, counterinsurgency situation, in which most of the supported ground force is from another nation. As in some of the other vignettes, many of the air operations would have to be long-range. There would be considerable need for surveillance and reconnaissance capabilities operating in a heavy-jungle environment.

Support for Myanmar

Myanmar (formerly Burma) is growing in strategic importance. Located between China and India, the nation has only recently started to come out of a period of intense, self-imposed isolation. China is starting to invest heavily in the nation, including building new major roads to facilitate access to the port of Rangoon in the south. This vignette postulates that a new group of leaders assumes control and tilts the nation toward the West and India, much to the dislike of the Chinese. In response, the Chinese back a revolt in which the pro-Chinese elements in the nation attempt to regain power. This revolt leads to an intense civil war. India expresses alarm at the situation and asks the United States to join its efforts to assist the Myanmar government. The United States elects to support the government with military advisers, supplies, and equipment, as well as selected air support. As in the Colombian vignette, regional and domestic political considerations mean that the United States must minimize its presence ashore—a reality that places a premium on carrier-based aviation.

Similar to several other vignettes, the distances to be covered by aircraft operating off the carrier would be long. Much of the fighting between government troops and the rebels takes place in the middle of the country, thus requiring air missions of 500 miles or more from carriers operating in the northern portion of the Bay of Bengal. ISR support for the government forces, plus occasional precision-strike missions, would be the main role of the carrier air wings in this situation.

Major Insights: Combat Vignettes

Insights were developed based on the entire range of possible combat situations—from very-high-intensity combat that included the possible use of nuclear weapons (China, Iran, North Korea) to less-threatening situations in which the carrier would be at relatively little risk (Myanmar). Likewise, they could be applied to most or all situations. The most important observations that COG-1 made are highlighted in the following subsections.

Reconfigure Carrier Air Wings

Each crisis is, of course, different. In some situations, the existing carrier air wing is appropriate for the situation. In other situations, the mix of aircraft aboard the ship will need to be modified. As discussed in Chapter Two, today's normal carrier air wing of roughly 70 aircraft includes a mix of aircraft types that can perform strike, non-stealthy reconnaissance, tanking, and anti–surface warfare and anti–submarine warfare missions. It is already an air element that has a useful mix of capabilities. Nevertheless, the current air wing is heavily weighted toward strike and anti-air operations.

Depending on the situation, the carrier will need to alter the mix of aircraft aboard, which may involve bringing non-naval aircraft onto the ship. The historical cases illustrated in Chapter Three show that doing so is not a new concept, inasmuch as non-naval aircraft have been operating from U.S. carriers since at least 1942. When a "standard" air wing might have to be modified to fit the specifics of the situation, most of the new aircraft flown aboard would be from the Department of the Navy—some mix of Navy and Marine aircraft. Occasionally, Army, SOF, or Air Force aircraft (either manned or unmanned) may have to come on board, but such situations will certainly be the exception, not the rule.

The important point is that, depending on the situation, the normal mix of aircraft might have to be altered, possibly on short notice, and even after the carrier has reached its operational area. Examples include the following:

- **Increasing the number of helicopters and/or vertical take-off aircraft (V-22s) on the ship, thereby providing a greater vertical-lift capability.** The COG determined this capability to be important in the Nigerian NEO, the counterinsurgency in Colombia, the Myanmar situation, and the Pakistani crisis. In some cases, the carrier would probably have to bring SOF personnel aboard and serve as their base of operations, at least temporarily, in addition to having to possibly evacuate military or civilian personnel back to the ship.
- **Increasing the number of reconnaissance and surveillance platforms of a different type.** Such platforms could be naval systems or, possibly, systems from the other services or Special Operations Command. Today, the Navy is investigating several unmanned aerial platforms that could carry various sensor packages. Since shipboard compatibility is a critical consideration for the Navy, these systems may have to be designed to meet naval constraints, thus possibly limiting the use of a common joint system that might be better suited for operations from large bases ashore.

- **Increasing the number of strike or air superiority aircraft.** In certain situations, the 50 strike aircraft aboard the ship may not be sufficient. The China-Taiwan and North Korea situations both provide examples of the need to increase the number of strike/air superiority–type aircraft.

To accommodate a different mix of aircraft, carriers would need to provide appropriate maintenance facilities, weapons storage, and berthing—which is particularly challenging if the aircraft are not from the Department of the Navy. Even when, for example, an unusually large number of extra Navy or Marine Corps helicopters join a carrier, the ship would need to alter its mix of spare parts and other key support items, which brings up the need for increased modularity.

Increase Modularity

Being able to modify the normal air wing aboard an aircraft carrier calls for increased modularity, which would entail being able to bring new capabilities, in the appropriate mix and in the right quantities, aboard the ship for specific missions. In this context, *modularity* refers to the ability to rapidly reconfigure the aircraft carrier for different, possibly nontraditional, missions. For example, in the Haitian intervention in 1994, two Navy aircraft carriers brought Army, Marine Corps, and SOF to the island. Had there been significant resistance ashore, the ships would probably have had to serve as bases of operations for those forces for several weeks. Today, aircraft carriers can certainly take aboard personnel and aircraft for nontraditional missions, but they are not well suited to act as a base of operations for nontraditional capabilities for extended periods of time.

Modularity would mean that the Navy would work together with the other services to make carriers better suited to supporting nontraditional missions and capabilities. The following are examples of modularity:

- Containers of spare parts and key maintenance equipment that can be brought on board the ship to support a greater number of Navy or Marine Corps vertical-lift aircraft (helicopters or V-22s) or aircraft of other services.
- Temporary, modular spaces for use by SOF elements that could deploy aboard ship for extended periods. This modularization should include the capability to insert, provide fire support for (as the gunship does), and recover SOF personnel.
- Modular medical facilities that would increase the ship's organic medical capability.

In many cases, the modules could be containerlike units that would be compatible with the ship's elevators and able to be stored in the hangar bay. Ideally, such modules would be fitted to the ship while it was in port, prior to a deployment. However, the likelihood that a crisis requiring a reconfiguration could occur after a carrier has entered, or is en route to, its normal operational area would require workarounds: Modules could be sent (by air or sea) to an intermediate staging base (e.g., Diego Garcia) for loading onto the ship. Unneeded portions of the normal air wing would be off-loaded to the intermediate staging base, together with unneeded personnel. It might also be possible to permanently store certain modules in forward locations, thus ensuring that they would be closer to key regions.

The COG-1 study team determined that existing ships would have the greatest difficulty in increasing modularity, since their designs are, of course, set. Future vessels, however, could be designed and built specifically to have improved modularity.

Enhance Reconnaissance and Surveillance Capability
This insight arose in every vignette the COG-1 examined. The desire for increased situational awareness is a crucial underlying concept for future U.S. military operations (whether their intensity is high or low). U.S. commanders at all levels want more information about the status of their own forces, the locations of noncombatants, and as much information as possible about the enemy. Therefore, a premium is being placed on increasing the number, type, and capabilities of ISR

systems, both manned and unmanned. An aircraft carrier can make an important contribution by providing commanders with persistent ISR capabilities for stealthy operation under cloud cover and with foliage-penetration radar (FOPEN) capability.

Today, a carrier has F/A-18 SHARP (strike-configured C/D and E/F variants also have ATFLIR targeting pods) (Navy) or LITENING targeting pods (U.S. Marine Corps [USMC]), E2C Hawkeye, and EA-6B Prowler aircraft. Each platform makes an important contribution to situational awareness. However, what a carrier lacks today is the ability to (1) project ISR at a very long range from land bases, (2) sustain ISR platforms at long distances (500–1,000 nmi), (3) collect tactical communications intelligence (COMINT), and (4) provide full-motion video. These latter two shortfalls result from the capability's having been either phased out of the carrier's arsenal (the ES-3's COMINT capability) or not included on the F/A-18, even though the technology is available.[1]

The need for greater long-range, long-endurance, under-weather, stealthy, armed, and unarmed ISR capability came to the forefront in each vignette that COG-1 examined. If carriers had the ability to project and sustain, to *at least 500 nmi*, a persistent (and synaptic over the regions of interest) ISR capability that included a mix of sensors (IMINT—electro-optical, radar, and other—and SIGINT, both COMINT and electronic intelligence [ELINT]), and the ability to quickly process and disseminate those data, the entire joint force would benefit. Specific examples would include the capability to

- quickly establish an "ISR blanket" in the vicinity of known or suspected enemy weapons of mass destruction (WMD) sites and possible missile-launch areas
- locate friendly and enemy forces and noncombatants in many different types of terrain

----

[1]   In the case of full-motion video reconnaissance, the U.S. Marine Corps has been using the low-altitude navigation and targeting infrared for night (LANTIRN) pod on its Harriers, but is now going to the improved LITENING pod.

- locate key enemy conventional capabilities, such as coast defenses, radars, air defense systems, and command nodes.

Increase the Range and Endurance for Covering Large Operational Areas

Many of the vignettes (Nigeria, Pakistan, Iran, Myanmar, Colombia) highlighted the fact that aircraft from a carrier, whether manned or unmanned, would have to operate at a great range (over 500 nmi) from the ship. This insight is supported by recent operations, such as Operation Enduring Freedom in Afghanistan during 2001–2002, when Navy aircraft ranged far inland on combat and patrol missions. Until Air Force and Marine Corps aircraft could start operating ashore in adequate numbers (a process that required weeks of political negotiations and considerable logistical preparation), aircraft carriers had to provide the overwhelming majority of tactical aircraft in the operational area. Many sorties flown by aircraft from carriers operating in the north Arabian Sea were hundreds of miles inland and of many hours duration. They overwhelmed the organic tanking capability of the on-station carriers, and large Air Force tankers had to supplement the ship's own tankers. The vignettes that COG-1 examined confirmed the possibility of future carrier operations taking place at great range from the ship.

The great distances that carrier-based aircraft would have to fly complicate the need for persistent coverage in the operational area. Being able to fly a long distance, drop ordnance, and return after spending only a short time in the target area may be appropriate in some situations. In others, being able to loiter over the area is highly desirable, either for ISR or strike purposes. Today's carrier air wing would have considerable difficulty maintaining more than a handful of platforms at distances of 500 nmi or more from the ship.

Prepare for Operations in a Nuclear Environment

Three vignettes (China-Taiwan, Iran, Korea) presented the possibility of enemy use of nuclear weapons, whether an overtly lethal and destructive attack by surface or aerial detonation or using a high-altitude nuclear detonation to disrupt U.S. C4ISR systems. Although either use would pose a great challenge to U.S. forces, naval forces, including aircraft carriers, have some distinct advantages and, possibly, some important vulnerabilities.

Most potential nuclear-armed opponents would have a limited ability to locate and track moving U.S. naval forces operating off their coasts. A Carrier Strike Group moving several hundred miles per day, 100 nmi off the enemy coast, would be very difficult for most opponents to locate, much less target. The carrier and its escorts would be able to detect and destroy most enemy reconnaissance systems, especially if they were airborne or surface craft. Enemy submarines might be somewhat more survivable. Civilian-type surface craft (e.g., fishing boats or merchant ships) may actually provide the enemy with the best means to locate and track a moving CSG, because the U.S. force would be less likely to attack them without considerable confirmation of hostile intent. Even when a supposedly innocent fishing vessel is providing periodic updates on the CSG's location, the targeting challenge would still be considerable: It might take several hours to program and launch a nuclear-armed missile toward the carrier's location. Compared with fixed land bases, a carrier is highly survivable, owing to its mobility and to the limited ISR capabilities that most opponents possess.

China would, however, have much better naval ISR capabilities than Iran or North Korea. China's ISR suite would include satellites and other systems that would substantially improve its ability to locate and track moving naval forces.

Although the carrier's mobility would complicate an enemy's ability to target it at sea, a nuclear-armed opponent may actually be more willing to attempt a nuclear strike against U.S. naval forces than against a fixed land base. Any nuclear strike against a U.S. or coalition base ashore would almost certainly cause considerable collateral damage and, possibly, massive numbers of civilian casualties. This reality might cause an opponent to be more hesitant in using a nuclear weapon against a shore target, despite the much greater ease of targeting a fixed facility. Meanwhile, a carrier that operates out at sea could be seen as a "purely military" target, and the destruction or disabling of the carrier could probably be done without many, if any, civilian casualties and other collateral damage. As mentioned above, the main challenge for most opponents would be locating and tracking a moving carrier and its escorts.

Another aspect of engaging a nuclear-armed opponent is the need for improved missile-defense capabilities. For example, the likely missile-launch areas of some of the possible opponents that the United States might have to confront in the future are within range of carrier-based capabilities. Improving the carrier's ability to detect and monitor missile-launch locations and to quickly engage missiles when they are launched would be a strategically important capability. For example, if carrier-based aircraft were armed with a boost-phase intercept capability to engage hostile missiles immediately after launch, an opponent's offensive missile threat could be significantly reduced.

CHAPTER FIVE

Uses of Aircraft Carriers in Future Noncombat Operations

The preceding chapter focused on the role of aircraft carriers in future combat operations. This chapter examines the use of carriers in non-combat situations. Most of these situations are humanitarian missions, with a focus on supporting domestic U.S. authorities in an emergency situation. As in Chapter Four, we provide a short overview of the vignettes, followed by the major insights that COG-2 derived. More details on each vignette are available in the Appendix.

Noncombat Vignettes

Nuclear Detonation in Long Beach Harbor

This is the most violent of all the vignettes involving support to domestic authorities. The case assumes that a radical nonstate terrorist group has managed to obtain a nuclear weapon and has smuggled it into Long Beach, California, aboard a container ship. The device is carefully concealed within one of hundreds of containers and is detonated before U.S. customs officials have an opportunity to inspect the ship. The weapon is about 10 kilotons, smaller than either of the atomic weapons that were used against Japan in 1945. Nevertheless, it is still capable of inflicting massive damage and releasing large amounts of fallout—especially since the explosion is a surface detonation.

The explosion inflicts considerable damage to the immediate vicinity of Long Beach Harbor. Massive fires are started in the port area, not just from the blast itself but also from the fuel escaping from wrecked ships. A few thousand people are killed or are seriously injured by the immediate blast. More ominously, fallout from the surface burst starts to arc northeastward into the Los Angeles basin, pushed by the prevailing winds.

There would, of course, be a need for a massive influx of relief efforts toward the Los Angeles area. Complicating the relief efforts would be the growing area of contamination that would gradually extend over more and more of Southern California. Literally millions of people attempt to flee the area. The U.S. Navy would certainly be called on to provide as much assistance as possible, including the capabilities from aircraft carriers. Ships from Bremerton, Washington, and San Diego, California, would make sorties to assist in this effort.[1]

Atlantic Tsunami

In this vignette, a major underwater earthquake occurs in the mid-Atlantic. The west coasts of Spain, Portugal, and North Africa are all hit by major tidal waves, as are portions of the U.S. east coast. Worst hit is the area from Norfolk, Virginia, to Savannah, Georgia. In some places, waves of up to 40 feet come ashore, wrecking infrastructure along the coast. Some low-lying communities in the Carolinas are very badly hit. In some places, bridges and coastal roads are washed away, thus isolating damaged communities from overland assistance. Only helicopters or boats can reach the isolated areas.

The Navy would already have deployed as many ships as possible from its bases in the threatened areas (primarily Norfolk, Virginia, and Moorehead City, North Carolina). Because of the short notice, however, only ships in a fairly high state of readiness would be able to deploy in the two or three hours of warning that would be available. Other ships already at sea, as well as vessels from other locations farther north and south, would converge on the area to provide assistance.[2]

---

[1]   This vignette is based on research RAND conducted for the Department of Homeland Security in a project that examined the consequence of clandestine smuggling and subsequent detonation of a nuclear device in Long Beach Harbor.

[2]   See McGuire (2005) for an interesting article on the possibility of a tsunami event in the Atlantic region.

Massive Volcanic Eruption on the Island of Hawaii

There is always volcanic activity on the Hawaiian Islands. This vignette assumes that the volcano of Kilauea on the Big Island (Hawaii) erupts with great force, causing massive damage to major portions of the island.[3] Additionally, the volcano's plume of ash flows in a northwesterly direction, carried by the prevailing winds toward other islands along the Hawaiian chain. The plume covers local airports with anywhere from a few inches to two feet of ash. Although geologists provided enough warning to permit a partial evacuation to start before the major eruption, there are still several thousand deaths and serious injuries on the Big Island. Roughly half of the island is covered by thick layers of ash or fresh lava flows.

The Navy would send all available ships to the area to provide assistance. As in the Atlantic tsunami vignette, pockets of survivors would be isolated from overland assistance, placing a premium on vertical-lift aircraft and boats from offshore ships.

Earthquake Strikes San Francisco Bay Area

This vignette involves the situation that has long been feared in California—"the big one"—which could cause massive damage to the region. Here it is assumed that a massive earthquake strikes the San Francisco area with relatively little warning. Considerable damage is done to local infrastructure, and several thousand people are killed or injured. Power grids are disrupted, thus complicating relief efforts.

Unlike the Long Beach nuclear-explosion vignette, there is no contaminated fallout to complicate the relief effort, and most people attempt to stay in the vicinity of their homes, if possible. Nevertheless, collapsed roads and bridges leave groups of people along the coast cut off and isolated.

This vignette also involves a complicating factor: a concurrent security crisis in the western Pacific. Just days before the earthquake,

---

[3]   This scenario assumes that a geological change has occurred in Hawaii and to the volcano's internal configuration that could result in an explosive eruption. Usually, the Kilauea volcano, unlike Mount St. Helens, for example, tends to vent pressure at regular intervals, with no subsequent explosive eruption.

the situation on the Korean Peninsula takes a significant turn for the worse, as the North Koreans, aggressively responding to multinational sanctions, threaten war. This threat results in numerous Pacific Fleet vessels being quickly dispatched to Asia.

Those naval vessels that remain in the eastern Pacific converge in the San Francisco area to assist in the relief effort. Debate within the National Security Council arises over whether to divert aircraft carriers from the crisis in Korea to participate in the relief effort in California.

Cuban Refugee Crisis

This vignette takes place in a post–Fidel Castro Cuba. Shortly after Castro's death, a struggle of succession grips Cuba and a quasi-civil war erupts among various factions. Some groups, including most of the military, back Fidel's brother, Raoul, to lead the country. Other groups, including elements of the Army, attempt to resist Raoul's succession. The situation rapidly degenerates into intense violence in some cities. Within a few weeks of the outbreak of fighting, large numbers of Cuban civilians attempt to flee the country, mostly toward the southern United States. Literally tens of thousands are in small boats at any time, headed toward Florida and other parts of the U.S. coast. It is hurricane season, which complicates the matter, increasing the risk of a massive humanitarian disaster should the flimsy small craft be caught by a westward-headed hurricane.

The U.S. government elects to block the influx of refugees and directs the U.S. Navy to intercept the oncoming mass of refugees while providing emergency humanitarian assistance to those in the boats who are in most dire need. Meanwhile, the U.S. government is seeking other spots in the Caribbean area, such as the Bahamas, to relocate the refugees.

Major Insights

Alter the Aircraft Mix Aboard Ship

All of the noncombat vignettes reveal a need to significantly increase the number of vertical-lift aircraft, because rescue and relief efforts ashore would be in areas in which few, if any, airports are available for conventional takeoff and landing aircraft. A premium would be placed on helicopters and V-22s that could reach isolated areas, bringing in relief supplies and emergency responders, as well as evacuating badly injured personnel, most of whom would be civilians.

Depending on the availability of Navy and Marine Corps helicopters and V-22s, there might be a need to bring Army and/or Air Force aircraft aboard. Most fighters and other strike-related aircraft would have to be off-loaded to quickly make room for additional vertical-lift assets. If these are aircraft types not normally in the carrier's air wing (Marine Corps V-22s or Army UH-60s or CH-47s, for example), there could be a need to also bring aboard additional maintenance equipment, spare parts, and key maintenance personnel. This need would depend, of course, on the situation. If a base is nearby at which Army or Marine Corps aircraft can receive maintenance, fewer, or possibly none, of those aircraft would need to be moved to the ship. At a minimum, Naval commanders would need to free up deck and hangar bay space for far more vertical-lift aircraft than the aircraft carrier normally supports.

Provide a Command Center for Key Government Personnel or Agencies

Depending on the level of devastation ashore, some key civilian government personnel may move onto the ship. In the case of the Hawaiian volcano, for example, the local phone and power systems might be so badly disrupted that key officials (Federal Emergency Management Agency [FEMA] or local government leaders) might have to move to the ship temporarily. This move would give them a secure location, with excellent communications facilities, from which to direct the relief effort.

Provide Medical Facilities for Casualties Brought Back to the Ship

Most of the vignettes we examined involved massive numbers of civilian casualties. The local medical facilities would almost certainly be overwhelmed, at least initially, until more capacity could be brought to the area and some of the casualties taken to other hospitals in areas not affected by the disaster. Until that is done, the Navy's ship-based medical capabilities could be of considerable assistance.

In such circumstances, it may be necessary to provide a modular capability aboard the ship. For example, to augment the normal medical facilities aboard ship, modularized medical containers and additional medical personnel could be flown to the ship and moved into the hangar bay.

The COG team noted, however, that aircraft carriers are not optimal locations for keeping patients for extended periods. Rather, carriers' medical facilities (augmented or not) are best used to provide immediate assistance to those seriously injured victims who would be brought back to the ship by the vertical-lift aircraft providing assistance to damaged areas. Indeed, in most cases, it would be better if the casualties, when located ashore, were taken directly to another shore-based medical facility.

In the nuclear attack on Long Beach, a major consideration would be the possibility that personnel and aircraft from the affected area could contaminate the aircraft carrier. Therefore, the radiation levels of aircraft crews, and casualties, returning from the devastated area would need to be monitored. An isolation area might be required, and if an aircraft is found to be highly contaminated by fallout, it might have to be pushed off the side of the ship.

Improve Availability of Nonready Carriers

The noncombat vignettes suggest that aircraft carriers could make a significant contribution to initial relief efforts. But as days pass and more civilian and other military capabilities are brought to bear in the area of the disaster, the role of the carrier likely would decrease. The ship's main value is in the first few days of the crisis. In this regard, the Navy should consider ways to improve the short-notice deployability of ships that are in port, in a semi-ready status.

Such ships would be those recently returned from deployments or having just completed yard periods. Although not ready for combat operations, these ships might be able to put to sea on fairly short notice (in one to three days, for example) to participate in a disaster relief effort. Time permitting, appropriate modules could be brought aboard to augment the ship's normal medical facilities or to provide maintenance for additional vertical-lift capability, particularly if the aircraft will be non-Navy or non-Marine.

Hold Carriers Back from Humanitarian Noncombat Missions When a Major Military Crisis Looms

The San Francisco earthquake vignette provides an example of a major disaster taking place simultaneously with a military crisis. Given the limited number of aircraft carriers that would be available for short-notice missions, it is likely that the carrier would be best employed for its primary mission: combat operations. Although the magnitude of the disaster and the availability of other military assets in the threatened region would be key variables, it appears likely that the senior U.S. military and political leadership would want to focus the carriers on combat. However, the preceding recommendation—improved availability of noncombat-capable carriers to respond to a disaster— would allow the Navy to provide more options to senior civilian leaders, who would want as many military capabilities as possible available if a major disaster occurred.

CHAPTER SIX
Conclusions

This monograph has examined how the Navy's aircraft carriers could be used in the future in missions ranging from high-intensity combat to humanitarian assistance. We provided historical examples of how carriers have been used in past nontraditional missions to show that these large ships have long been able to provide capabilities beyond the air superiority and strike roles that are traditionally associated with the aircraft carrier.

The vignettes that we examined in this research spanned the range of military operations. In some highly challenging situations, aircraft carriers might have to conduct operations under the threat of a possible enemy nuclear-weapons use. In other cases, the carrier's traditional strengths of strike and air warfare would be the main capability that the ship would provide. In so-called low-intensity operations, the carrier's large flight deck could be used for aircraft performing surveillance and reconnaissance functions. Finally, the ability to alter the aircraft mix aboard the ship could be very useful for humanitarian operations, particularly when many vertical-lift aircraft would be needed.

The entire series of vignettes—combat and noncombat—highlighted the carrier's value in crisis management. Because these ships can often be the first significant U.S. military capability to arrive in the area of a crisis, they can be used to immediately develop a better understanding of the situation by providing various reconnaissance and strike capabilities. Additionally, the carrier can serve as a platform for non–Department of the Navy (DoN) capabilities (i.e., Army) or for SOF elements that might otherwise be limited in their ability to gain access or base near a crisis point.

No current or projected manned or unmanned, sea- or land-based platform provides the capabilities needed in the most severe crises examined.

Throughout the examination of the vignettes, several recurring insights came to the forefront. First was the need for flexibility, especially for altering the aircraft mix aboard the ship. In combat-oriented situations, there could be a need to add more reconnaissance and surveillance capability (via more manned or unmanned aircraft) to support naval or joint operations. In humanitarian-assistance or low-intensity operations, there could be a need for many more vertical-lift aircraft aboard the ship than are normally present in a Carrier Air Group. Importantly, the need to alter the aircraft mix aboard the ship could occur while the carrier is on a normal deployment, in response to an unexpected crisis, thus requiring a short notice and significant change in the aircraft mix. Another possible reason for an alteration of the aircraft mix is a preplanned operation for which different aircraft types—more DoN aircraft (probably from the Marine Corps) or Army aircraft—could be brought aboard the carrier. Historical precedent shows that carriers have been used for this purpose ever since the Doolittle Raid of April 1942 and as recently as USS *Kitty Hawk*'s serving as a platform for Special Operations aircraft and troops near Afghanistan in 2001–2002.

The need for flexibility requires a more modular approach. Current ships could be reconfigured, but at some cost, to increase their ability, for example, to support a different aircraft mix. Future ships, particularly the soon-to-be-built CVN-21, could be modified while still in the design phase to increase their ability to switch missions and aircraft types, realizing that air warfare and strike will still be the primary missions of the aircraft carrier.

It should be noted that representatives from Britain's Royal Navy participated in this research, offering insights from Royal Navy operations, as well as information on the future 65,000-ton Royal Navy aircraft carriers (CVF), which will be deployed in 2012–2015. In keeping with the increased emphasis on joint operations in the British military, the two CVF ships are being designed specifically to accommodate—at least occasionally—British Army and Royal Air Force aircraft. In keeping with the current flexible use of today's 20,000-ton Invincible-class carriers (which have on occasion transported battalion-sized elements of Royal Marine commandos), the CVFs are envisioned as being able to conduct various types of joint operations in addition to performing their traditional role as strike and air warfare platforms.[1]

The second major, recurring insight that came up in virtually all the vignettes was the need to enhance the aircraft carrier's reconnaissance and surveillance capability. To be sure, this is more a function of the type of aircraft carried onboard the ship (manned and unmanned) rather than ship design per se. Nevertheless, one of the steps needed to improve the carrier's versatility as a base for reconnaissance and surveillance assets is a change in thinking about how the ship will be used in future operations. In many future operations, especially low-intensity missions (which are far more frequent than major combat), senior U.S. naval and joint commanders will want high-quality surveillance and reconnaissance capability in order to improve situational awareness.

One of the major themes of most future U.S. military concepts is the desire to have a major advantage in situational awareness over the nation's opponents. Precise, real-time knowledge of the locations of friendly forces, noncombatants, and the enemy are a key goal of today's and the future's U.S. commanders. Increasing the carrier's ability to contribute to situational awareness by providing it with a wider suite of reconnaissance and surveillance platforms, together with improved on-ship processing capability, could be a significant addition to the carrier's ability to support the entire joint force, particularly with respect to the potential employment of nuclear weapons carried by ballistic or cruise missiles: It is crucial to be able to detect key events, such as the movement of nuclear warheads out of storage sites, and to detect preparations for missile launch. Such detection will require that high-resolution imagery be available, without significant delay after the need occurs, and be able to cover all storage and launch sites, all the time, regardless of cloud cover.

---

[1] The concept of increased modularity and flexibility for current and future U.S. Navy aircraft carriers could also be applied to the next large amphibious ships—the LHA-R class, which is intended primarily to serve as a platform for Marine Corps operations.

Finally, most of the vignettes highlighted that carrier-based assets (whether strike, air superiority, or reconnaissance/surveillance) need to have longer ranges. In the future, the Navy should strive to provide carrier-based platforms with a range of at least 500 nmi, as well as improved loiter and endurance capabilities. Today, the Naval Air Systems Command (NAVAIR) is exploring options for future "Persistent Unmanned Airborne Surveillance Capability," a research effort that should consider ways to provide carriers with (1) several unmanned reconnaissance platforms with a mix of sensors to provide reconnaissance and surveillance in different terrains and (2) systems with the range and endurance to operate at great distances from the ship for 12 hours or more.

Changing the aircraft mix aboard a carrier (even if only temporarily), increasing the ship's modularity, and placing more emphasis on reconnaissance and surveillance all entail trade-offs. As large as a modern aircraft carrier is, it still has a finite amount of deck and hangar space. Increasing onboard processing capability of raw sensor data will take up space and compartments that are currently devoted to other purposes. Importantly, the carrier's forte for over 65 years has been its ability to conduct air superiority and strike operations. It is likely that these will remain the primary roles of the carrier for many years to come. Therefore, alterations in the ship's role mean that less deck, hangar, and other space will be available to support aircraft devoted to strike and air warfare.

That said, the U.S. military as a whole (including the Navy) faces many nontraditional challenges that could last many years into the future. Increasing the aircraft carrier's flexibility will therefore benefit the entire joint force and the options available to senior U.S. military and civilian decisionmakers.

To summarize, the major insights developed in this study include the following:

- Combat recommendations:

1. Improve abilities to reconfigure carrier air wings.
2. Increase carrier modularity.

3. Obtain greater reconnaissance and surveillance capabilities.
4. Increase the ability of carrier aircraft (manned and unmanned) to operate at greater range and endurance over large operational areas.
5. Prepare for operations in a nuclear environment.

- Noncombat recommendations:

1. Enhance carrier abilities to alter the aircraft mix aboard ship.
2. Enhance carrier abilities to provide a command center for key government personnel or agencies.
3. Enhance carrier abilities to provide medical facilities for casualties brought back to the ship.
4. Improve the availability of nonready carriers for responding to unforeseen crises.
5. Hold carriers back from humanitarian noncombat missions when a major military crisis looms.

It should be noted that this research did not specifically examine the Navy's large amphibious ships: LHAs and LHDs, ships designed to support Marine Corps expeditionary forces. Although they do have large flight decks that can support helicopters, tilt-rotors, and some fixed-wing attack aircraft, they lack the ability to launch and recover high-performance strike aircraft. Nevertheless, some of the insights developed in this research might be applicable to LHAs and LHDs. For example, the ability to quickly alter the aircraft mix aboard ship in response to unforeseen circumstances could also apply to the large amphibious ships.

This research did not include all possible situations in which aircraft carriers could be employed in the future. The study nevertheless does cover a wide range of representative cases in which one or more aircraft carriers could be employed. The examined vignettes were intended to represent a broad range of hypothetical operations, ranging from high-intensity combat—including the possible use of nuclear weapons—to noncombat humanitarian situations. Various regions were deliberately included in the analysis in order to explore the ramifications of aircraft carriers operating in open ocean areas (such as off the coast of China) to more-constricted areas, such as the Persian Gulf and North Arabian Sea.

This study has highlighted how aircraft carriers might be used in a wide range of future operations. As the Navy considers how these ships will be used in the future, and the types of aircraft that will operate from their decks, it is hoped that this and other studies will give the service useful insights into how these already-multirole ships can become even more useful.

Future Combat and Noncombat Vignettes

This appendix details the scenarios, or vignettes, that the RAND research team and the Concept Option Groups used to examine how and under what conditions aircraft carriers might be used in the future. COG-1, the group that investigated how aircraft carriers might be used in future combat operations, considered seven vignettes; COG-2, which explored uses of the vessels in future noncombat situations, looked into five vignettes.

Future Combat Operation Vignettes

China-Taiwan Crisis (2009)

Background. Since early in this decade, China has conducted a buildup of its conventional military capability. The goal has been to have a viable military option to use against Taiwan, in an attempt to coerce the island into joining with the mainland. To have a usable military option, China has also had to improve the capabilities it would need to deter or to actually fight against the United States.

The Chinese have devoted most of their efforts to increasing their air and naval capabilities. New attack submarines (diesel and nuclear) have been added to the fleet, as have a considerable number of amphibious vessels. The People's Liberation Army–Air Force has added Russian MiG-31 and Su-27 fighters, domestic J-10s, and several battalions of SA-10, SA-12, and SA-20 surface-to-air missiles (SAMs). The SA-20 can engage targets over Taiwan itself from firing locations in eastern China. The 2nd Artillery Corps (missiles) is now armed with over 1,000 medium-range ballistic missiles, some of which are accurate to within 100 meters. Most of these missiles can reach the U.S. airbases on Okinawa. More than 100 long-range missiles capable of reaching Guam are now available. They have conventional or nuclear warheads. Selected Chinese ground units (roughly six divisions) have been extensively modernized with new equipment. Finally, a substantial amount of effort has been devoted to information warfare and reconnaissance—in particular, to developing surveillance and reconnaissance systems capable of locating U.S. naval forces.

The government in Beijing waited until after the 2008 Olympic Games before taking a direct coercive approach against Taiwan. The motivation to take such direct action increased during the games, since the leadership in Taipei used China's refusal to admit its team into the games as an opportunity to pronounce its "new national status" and virtually declare independence from the mainland. Within a month of the end of the games, Beijing announced major military exercises that included Chinese warships operating north and east of the island. On the political front, Beijing issued a virtual ultimatum to Taipei that it must renounce any pretense of independence and enter into bilateral talks to "formalize the union of Taiwan with its home country." The leadership in Taiwan refused to do so.

Scenario: January 2009. On January 15, Chinese warships seized two Taiwanese freighters that were en route to the island, escorting them to Chinese ports. Both ships were loaded with military equipment just purchased from Europe. Once the cargoes were inspected, Beijing announced that the weapons represented an outright provocation and declared an economic blockade of the island. Chinese warships put to sea, and the entire Chinese military started mobilization. The government in Taipei also announced mobilization, and it called on the United States, the United Nations, and other nations for assistance to repel Chinese aggression.

The President directed the Secretary of Defense to dispatch several Carrier Strike Groups, additional long-range bombers and support aircraft, Army missile defense units, and two Marine Expeditionary Units (MEUs) toward locations in the western Pacific, although not to Taiwan itself.

As soon as these steps became known, the Chinese government declared them an attempt to interfere with internal Chinese affairs. Ominously, Beijing's ambassador in Tokyo handed a formal communiqué to the Japanese government, warning it to remain strictly neutral in the event of an armed conflict over Taiwan and stating that any American use of Japanese bases for operations against China would be regarded as an act of war by Japan against China. The communiqué stated that China would feel free to take any appropriate steps in those circumstances.

On January 20, roughly three days into the U.S. deployment of additional forces toward the western Pacific, the Chinese commenced military operations against Taiwan. A barrage of over 200 conventionally armed missiles strike Taiwanese air bases, command centers, and key political locations. Less than 30 minutes later, a strike by some 250 Chinese fighters and bombers hits military targets on the island. Later that day, a Taiwanese freighter some 200 miles off the east coast of the island is torpedoed and sunk, apparently by a Chinese submarine. Considerable movement of coastal vessels has been detected in Chinese ports close to the straits between Taiwan and the mainland.

That afternoon, the President authorizes U.S. military action to defeat Chinese attempts to isolate or take over Taiwan. Almost simultaneously, the Japanese ambassador in Washington informs the Secretary of State that U.S. military forces located in Japan are not allowed to conduct offensive operations against China until the Japanese Diet considers the matter. He informs the Secretary of State that Tokyo has just received what amounts to a formal ultimatum from Beijing, demanding that Japan prohibit any U.S. use of its bases.

Coup Attempt in Pakistan (Year Unknown)

Background and Scenario. Radical elements in the Pakistani armed forces have staged a coup in an attempt to overthrow the government. Although the coup failed—the government of Pakistan remained in power—the rebels managed to seize control of two Pakistani nuclear-weapons storage sites, together with a number of missile launchers. A number of Pakistani military units have joined the rebels and are now defending the seized nuclear facilities. Recognizing the gravity of the situation, the Pakistani government has requested U.S. military assistance in order to (1) defeat the rebel forces, (2) prevent the launch of a nuclear weapon, and (3) prevent a nuclear weapon from being removed from the storage sites and delivered to parties unknown.

The following are details of the scenario:

- *Rebel forces:* Several brigades of the Pakistani army have sided with the leaders of the coup attempt. Both of the nuclear-storage facilities that are in rebel hands are currently defended by brigade-sized rebel forces (2,000–3,000 men each) equipped with armored vehicles, artillery, and some air defense systems, including anti-artillery (AA) guns, MANPADS (SA-16 and SA-18), and low-altitude beam-rider weapons (Swedish RBS-70). It is known that several Ghauri missile launchers are in the hands of rebel forces and are located in the general area of the nuclear-storage sites that are under rebel control. Although these launchers are currently well dispersed and hidden, they could be quickly brought to the area of the storage sites, unless they are blocked or destroyed by air or ground forces.

- *Nuclear sites:* Both of the nuclear-weapons storage sites under rebel control are located in rural areas and are hardened against attack. The Pakistani government is unwilling to provide detailed plans of the facilities but has told U.S. authorities that the facilities have several exit/entrance routes (which they are willing to identify) and are deeply buried. According to Pakistani military officials, they feel that no nonnuclear weapon can penetrate into the main portion of the underground facilities, although the entrances themselves are vulnerable.

- *Pakistani forces:* The Pakistani military has dispatched a division-sized force toward each of the nuclear-storage sites under rebel control. They have not yet made direct contact with the rebels. Pakistani air force elements have attempted to conduct reconnaissance of both sites and have been fired on by the rebels: Two Pakistani fighters were shot down over one site, apparently by RBS-70 beam-rider missiles.

- *Indian reaction:* The Indian government has expressed great alarm over the current situation. Significant portions of the Indian military have been placed on a high state of alert.

North Korea Crisis (2006)

Background and Scenario. By spring 2006, it is clear that the six-party (United States, Republic of Korea [ROK], Democratic People's Republic of Korea [DPRK], PRC, Russia, and Japan) negotiations to eliminate the North Korean nuclear-weapon programs have reached an impasse. With strong U.S. encouragement, China, Japan, and Russia announce the beginning of a "partial" economic and financial embargo on all trade with the DPRK—most importantly, including Chinese petroleum. Although officially supportive, South Korea expresses grave private doubts that the strategy of economic coercion will lead to a second Korean War. There are large demonstrations in South Korea, protesting the ROK government's tentative support for the coercive strategy. Pyongyang virulently denounces this turn of events and warns of war.

In response to the growing discontent in South Korea, the United States announces that it will take "prudent precautionary measures" to ensure that North Korea does not consider a military option to escape an "unavoidable decision to give up its nuclear-weapon program." Further, the Secretary of State announces that Washington would treat any evidence that the DPRK is selling and/or transferring nuclear weapons and/or weapons-grade material to parties "known or unknown" as a "threat to its vital interests." U.S. and Asia media interpret this statement as a "red line" for North Korea.

Consistent with the Proliferation Security Initiative, the United States and Japan declare a "naval surveillance zone" around the DPRK to ensure that Pyongyang does not attempt to transfer nuclear-weapon materials by sea.

The crisis rapidly escalates after three events. First, a U.S. Navy (USN) warship stops a DPRK freighter in the China Sea. The North Korean crew resists an attempt by the USN to board the freighter. During a short firefight, the North Korean ship blows up. Washington is very suspicious of a "war provocation." Second, North Korea responds by firing two FROG-7–class short-range ballistic missiles (SRBMs) into the Han River between Seoul and Inchon in an attempt to intimidate the South Korean government. Third, the Japanese Coast Guard has an armed clash with a North Korean "spy boat" off the west coast of Kyushu. Pyongyang responds by threatening to "demonstrate" its nuclear deterrent against Japan. During an interview with the South Korean media, the North Korean leader appeals for "Korean unity" and promises not to use his nuclear arsenal against his "Korean brothers and sisters." This interview creates a sensation among the South Korean public. There are mass demonstrations in the major cities of South Korea, calling for the withdrawal of all U.S. forces. The U.S. Ambassador to South Korea reports that the ROK may not allow U.S. reinforcements or the use of its airfields for offensive operations in the event of war. In turn, the Japanese Prime Minister, through a hotline, calls on the U.S. President to stand firm in the face of North Korean nuclear aggression.

The U.S. government is now considering its options to take action against North Korea, including the possibility of a preemptive strike against the North's WMD storage facilities and missile-launch units. The following are details on the scenario:

- *North Korean armed forces:* During fall 2006, the Intelligence Community (IC) concludes that the DPRK has between 12 and 18 plutonium weapons of the design provided by A.Q. Kahn. They are estimated to have a yield of 10 to 15 kilotons and capable of being fitted on the NoDong-class medium-range ballistic missile (MRBM), which can reach all major Japanese cities. Additional plutonium weapons are under production and may arm either additional NoDong MRBMs or a 600-km-range Ground-Launched Cruise Missile (GLCM) that had been repeatedly test-flown during summer 2006. Evidence of the status of the uranium-enrichment program remains murky, but the IC is convinced that a centrifuge array will be operational by the end of 2007 and will have the capacity to produce up to two 15-kiloton-class uranium bombs a year.

- *South Korean armed forces:* The South Korean armed forces are on a high state of alert following North Korea's demonstration shot into the Han River. South Korea has a very limited ability to intercept North Korean cruise missiles and is totally dependent on the United States for defense against ballistic missiles. The South Korean government has strongly indicated that it will not allow the United States to use its bases for offensive strikes against the North, unless there is another attack directly against the ROK.
- *Japanese armed forces:* Japan has a very limited ballistic-missile-defense capability, but a somewhat better ability to detect and intercept cruise missiles approaching over water. The Japanese Self-Defense Forces have increased both their alert level and their protective measures around their own bases, as well as around U.S. facilities in Japan.

Crisis with Iran (Year Unknown)

Background and Scenario. Since U.S. forces occupied Iraq in early 2003, the Islamic regime in Iran has stepped up its efforts to undermine and weaken U.S. power and influence in the Persian Gulf region. Realizing that they cannot directly challenge U.S. military power, the Iranian regime has used third parties—terrorist groups, radical organizations, and insurgents—in the Middle East to attack U.S. interests and undermine the governments of nations in the region that are friendly to the United States. Such indirect actions against the United States have also placed Tehran at some risk, because it cannot completely control the actions of its third-party proxy forces.

Meanwhile, Iran has become a nuclear-armed nation, although publicly denying that it has any nuclear arms. At present, it has roughly a dozen nuclear weapons. Delivery means include theater missiles that are capable of striking Israel, Turkey, and all of Saudi Arabia. Additionally, some Iranian aircraft have been refitted to drop nuclear bombs.

The precipitating event for the crisis is a terrorist bombing attack against a hotel in Amman, Jordan, that is frequented by Westerners.

The hotel was the scene of an international conference on aid to poor African and Middle Eastern nations. More than 400 people were killed in the blast, including roughly 100 Americans. Two of the terrorists were captured by Jordanian authorities. Upon interrogation, one states that his group developed the plan of attack and was armed by Iranian agents. Additional leads are followed, and the evidence pointing toward direct Iranian involvement seems undeniable.

When the United States publicly accuses Iran of sponsoring the bombing, the leaders in Tehran openly claim that they are prepared to respond "massively" to any U.S. attack, and they state that for any nation in the region to assist in or contribute to a U.S. attack on Iran will be regarded as an act of war. Immediately, several Middle Eastern governments that are generally friendly toward the United States request information on U.S. intentions toward Iran and information on what protection the United States can provide them if the crisis leads to actual armed conflict.

The following paragraphs detail this scenario:

- *Iranian forces:* In the past few years, the Iranians have used oil and gas revenues to (1) fund their clandestine nuclear program and (2) purchase a significantly increased air defense system to protect key locations in the country. The Iranian air force now includes several squadrons of MiG-31 and MiG-35 fighters and a small number of airborne early-warning aircraft. Additionally, the Iranians have deployed ten batteries of SA-10/SA-20–type SAMs. These batteries are concentrated around Tehran, the largest oil fields, the Straits of Hormuz, and the nation's nuclear infrastructure. Several batteries of SA-15 SAMs complement the longer-range SA-10s and SA-20s. The Iranian military has also placed great emphasis on commando operations and is known to have conducted exercises in the Persian Gulf, where civilian fishing vessels were used to carry commandos. As mentioned above, the Iranians have both a missile-delivery option and an aircraft-delivery option for their small nuclear arsenal. Finally, they have several hundred conventionally armed theater missiles that can range as far as Israel and western Turkey.

- *Nuclear sites:* Iran has a number of nuclear facilities. There are several suspected nuclear-weapons storage sites, but the exact location(s) of Iran's nuclear weapons has not been determined.
- *Iranian proxy forces:* Several Middle Eastern terrorist and insurgent groups are effectively Iranian proxies. Indeed, one of these groups conducted the bombing that has brought on the current crisis. These groups have the ability to strike military, political, and economic targets throughout the region, although their capability to penetrate a heavily guarded facility is very limited.
- *Reaction by nations in the region:* Most nations in the region are alarmed by the turn of events, especially since most have long suspected that the Iranians do have an operational nuclear arsenal. Particularly concerned are those nations that are openly friendly to the United States. Several of these countries have quickly sent messages to Washington, requesting positive assurances that the United States will protect them in the event of armed conflict.

Nigerian Civil War (2010)

Background and Scenario. For years, there has been marked tension between various ethnic, tribal, and religious groups in Nigeria. This country of some 123 million people comprises roughly 250 ethnic groups. Of these, ten major groups account for about 90 percent of the population. Approximately half of the population (mostly in the north) is Muslim; Christians and Animists dominate the southern region. Historic tensions and suspicions remain from the 1960s' bloody and bitter civil war, when the Ibo tribal group in the eastern portion of the country tried to break away to create the independent republic of Biafra. This era was followed by ever-increasingly corrupt military regimes. Since 1999, the country has had a democratically elected government. This move from an authoritarian military-dominated "kleptocracy" has had positive effects. However, the country's internal tensions have continued to rise, and there have been episodes of violence in the northern regions, mostly against those who oppose the increased use of Islamic law.

Even with higher sustained oil revenues during the first decade of the 21st century, the Nigerian government continues to demonstrate a high degree of inefficiency and corruption. In turn, the Muslim northern regions have taken an increasingly aggressive stance on their right to impose Islamic law. Further, the U.S. government has been warned by the British government that elements of Al Qaeda have successfully established a growing presence in the northern region. The central government has been unwilling and/or unable to crack down on this phenomenon. During summer 2010, Islamic militant groups launch a series of pogroms aimed at Christian and Animist populations in the north and central part of the country. Following one particularly bad urban massacre, the central government intervenes with military force, which results in widespread violence in the country. Most ominously, several large Nigerian Army units defect to the side of the Islam-inspired insurgents.

Very rapidly, the U.S. government and key allies—most specifically, the United Kingdom (UK)—face a large-scale NEO. Thousands of U.S. and European Union (EU) nationals are spread throughout the country. From a military perspective, the most demanding challenge is to find, collect, and evacuate civilians from deep in the country, some several hundred miles from the coast. It is in this region that the heaviest military fighting, including reports of the first atrocities against non-Nigerian nationals, is taking place.

The Nigerian government calls for military assistance from the United States and the EU to effect a series of NEOs and provide longer-term counterinsurgency assistance.

The following paragraphs present details on this crisis:

- *Rebel forces:* The Director of National Intelligence (DNI) reports to the U.S. President that the IC believes that armed Islamists number well over 10,000, mainly organized in platoon- and company-sized roving bands. Several brigades of the Nigerian Army, numbering some 15,000 troops, have defected to the Islamic insurgents. These three brigades are well equipped with some modern infantry and crew service weapons, which include precision-guided anti-aircraft and anti-armor weapons. The IC is unable to ascertain whether these units will actively oppose an NEO by either the United States or the European Union.

- *Nigerian forces:* The Nigerian armed forces are dominated by the Army, which is made up of some 150,000 regular troops. Most are moderately well equipped and well trained, but they are spread all over the country for internal security purposes. Lagos has three elite commando brigades for use in peace enforcement and stabilization missions. Currently, one of these brigades is tied up with a United Nations (UN) mission in Southern Sudan. The Nigerian Air Force is of modest size and capability. For this crisis, its most valuable assets are approximately 100 helicopters and light transport aircraft. The Nigerian Navy is basically a coast guard and not militarily relevant to this particular crisis.
- *EU forces:* The UK has the capacity to intervene within a matter of days with the equivalent of one light brigade that can be air-transported to Nigeria. A follow-on Marine brigade can arrive in the area within three weeks by sea, with a light carrier providing vertical takeoff and landing (VTOL) fighter cover. Other EU states, such as France, can provide similar resources: several light brigades that can be air-landed and one or two that can be deployed by sea within several weeks.

Colombian Insurgency (2010)

Background and Scenario. The Colombian government has been able to conduct a forceful counterinsurgency campaign during the early years of the first decade of the 21st century against its main ideological enemies, the FARC and ELN. Simultaneously, it has been able to largely pacify the various right-wing insurgent groups through a mixture of military coercion and amnesty programs.

By the end of the decade, however, these favorable trends have seriously deteriorated. The consolidation of power of the Hugo Chavez regime in Venezuela has provided both the FARC and ELN with the opportunity to obtain clandestine support and sanctuary from a sympathetic government. In addition to the direct and indirect assistance of the Venezuelan government, the FARC, in particular, has been able to generate sufficient revenue from the illegal drug trade to purchase advanced infantry weapons on the world market. Most worrisome has been the increasingly widespread and effective use of advanced (SA-16–type and SA-18–type) MANPADS, "smart" mines, and man-portable precision weapons, such as anti-tank guided missiles. These acquisitions have done much to counter Plan Colombia, the sustained modernization of the Colombian armed forces, especially its acquisition of a helicopter fleet to provide it with tactical and operational mobility. Additionally, the insurgent organizations have been able to acquire more-sophisticated communications systems to enhance their own command and control capabilities.

Although the United States is committed to supporting the Colombian government, it remains preoccupied by the challenge of effecting a successful conclusion to the Iraq War and other global national-security priorities in the Greater Middle East region. It is during this time that both the ELN and FARC embark upon a series of aggressive campaigns to regain control of many rural regions of Colombia that heretofore have been under government control. During spring 2010, the FARC makes a strategic decision to launch a major campaign inside the major cities, including Bogotá. This Tet-style offensive has stunning initial military success and prompts the Colombian government to call for direct U.S. military intervention—not only to rescue the government from the immediate military emergency but also to support a vigorous strategic counterattack against FARC and ELN bastions that have re-emerged in various rural zones.

Although several neighboring states have expressed sympathy with the Colombian government's plight, Ecuador, Brazil, and Peru strongly oppose any dramatic military intervention by the United States. Citing its peacekeeping experience in Haiti, the Brazilian government offers direct military assistance as an alternative to a U.S. military escalation. Intense negotiations begin between the United States, Colombia, and Brazil on whether a militarily viable "coalition of the willing" can be constructed to save the Colombian government from a military disaster. Throughout this period of crisis, the Venezuelan government both proclaims its innocence and denounces any military intervention by the United States.

The following paragraphs provide details on the key players in this scenario:

- *Rebel forces:* The DNI provides an IC estimate to the U.S. President that the FARC is now organized into ten well-equipped brigade-sized "operational groups" of 3,000 men and women each. The FARC has an additional 40,000 "militia" providing rear area support. The best estimate from the IC is that the ELN has a smaller force of approximately 30,000 men and women organized into small battalion-sized units. Both are well equipped with the full range of small arms including light and heavy machine guns, mortars, and high-performance sniper rifles. As noted above, both organizations have been able to obtain and use effectively an array of precision-guided anti-aircraft and anti-armor weapons.

- *Colombian armed forces:* The Colombian armed forces number some 200,000 men and women. It has been rather well trained in counterinsurgency operations and has recently behaved better toward its citizens. The Colombia Air Force now operates the largest fleet of helicopters, some 250 aircraft, in Latin American. Unfortunately, both the Colombian Army and Air Force have suffered severe losses during the on going "Tet" launched by both the FARC and ELN.

- *Brazilian armed forces:* The Brazilian government has offered to deploy a brigade of elite commando and infantry units to help stabilize the military situation in Bogotá. It has requested that the United States provide airlift assistance.

- *Venezuelan armed forces:* The Venezuelan Army stands at 70,000 active forces, primarily of light and motorized infantry. It is "reinforced" by a 400,000-person "Bolivarian Popular Militia." Even with substantial and sustained higher oil revenues during the early 21st century, the Chavez government provides the Venezuelan Air Force and Navy with modest modernization programs. Both remain under suspicion of being "anti-regime." There is considerable intelligence that shows that Venezuela is providing operating bases for the FARC and ELN within its territory. Various small camps close to the Venezuelan-Colombian border have been noted, and there is a strong indication that the Venezuelan army has passed weapons and other key items of equipment to both groups.

Support for Myanmar (Year Unknown)

Background and Scenario. Myanmar (formerly Burma) is growing in strategic importance. Located between China and India, the nation has only recently started to come out of a period of intense, self-imposed isolation. China is starting to invest heavily in the nation, including building new major roads to facilitate access to the port of Rangoon in southern Myanmar. This vignette postulates that a new group of leaders assumes control and tilts the nation more in the direction of the West and India, much to the dislike of the Chinese. In response, the Chinese back a revolt in which the pro-Chinese elements in the nation attempt to regain power, which leads to an intense civil war. India expresses alarm at the situation and asks the United States to join its efforts to assist the Myanmar government. The United States elects to support the new Myanmar government with military advisers, supplies, and equipment, as well as selected air support. As was the case in the Colombian vignette, regional and domestic political considerations mean that the United States must minimize its presence ashore in the area. This reality places a premium on carrier-based aviation.

Similar to several other vignettes, the distances to be covered by aircraft operating off the carrier were assumed to be long. Much of the fighting between government troops and the rebels is taking place in the middle of the country, thus requiring air missions with a radius of 500 miles or more from carriers operating in the northern portion of the Bay of Bengal. ISR support for the government forces, plus occasional precision-strike missions, would be the main role of the carrier air wings in this situation.

Future Noncombat Operation Vignettes

Long Beach Nuclear Detonation (2009)

Background. In early January 2009, an intelligence intercept led to the conclusion that a coalition of leaders among the disparate Chechen rebel groups had for some months been seeking a unifying goal that would help champion their independence cause and galvanize greater commitment and support both within Chechnya and from the Chechen diaspora elsewhere in Russia and globally. Acquisition of one or more nuclear weapons was reported to be among the strong contenders for this flagship galvanizing effort.

In late April 2009, Russian special units locked down a tactical nuclear-weapons storage facility east of the Urals. Reports from communications intercepts suggest that there was an attempted theft of nuclear weapons at the facility by a group of Chechens. The Russian government has subsequently rebuffed all efforts to confirm rumors of a "major nuclear theft" of as many as a half-dozen weapons at the facility.

In a separate development, National Security Administration (NSA) communications intercepts indicate that authorities at a Minatom nuclear facility have been discussing another theft, this time of weapons-grade, highly enriched uranium, which may have taken place more than a year ago. The quantity of material stolen may have been enough to manufacture as many as three or four crude nuclear weapons.

Scenario: The Nuclear Detonation/Monday, June 6, 2009/Long Beach Harbor—1130 PST (1430 EST). On a routine Monday morning, inspectors at the Port of Long Beach are responding to the latest regulations on enhanced container inspection for both incoming containers from overseas and containers to be shipped from Long Beach to ports farther north and in Asia.

Port of Long Beach—1200 Noon PST (1500 EST). At the Port of Long Beach, a nuclear weapon detonates in a tremendous explosion heard and seen throughout the Los Angeles Basin. Large quantities of materials and water are immediately sucked into a forming cloud of debris, and a large mushroom cloud begins to rise from the port area.

As a result of the explosion, there is damage to structures and flying debris injures people as far as 1.6 miles from the blast center. Physical damage to the city of Long Beach from the blast wave is relatively light (broken windows and flying debris, but roofs are mostly intact) and is limited to the southeast corner of town, but that area is a business center, full of office workers.

People within approximately 2.2 miles of the blast center who were not shielded by buildings or other structures suffer from flash burns over the portion of their bodies that is facing the port. The affected area reaches as far as Long Beach Plaza. The hillside community of San Pedro, due west of the blast site, receives disproportionately extensive damage because of its direct line of sight to the blast area. People within approximately 0.75 mile of the blast center who were not shielded absorb dangerously high prompt doses of radiation, although these effects are limited to the port area and miss the city. Those who were exposed to fatal levels of radiation may not develop symptoms for hours or even a few days. They may take days or even a few weeks to die.

As a result of the blast, containers from neighboring cargo ships are scattered at high velocity. Some ships suffer hull ruptures at the waterline on the side facing the explosion, including a crude-oil tanker that was off-loading crude at Pier T. Crude from the tanker begins to flow rapidly into the harbor and is soon ablaze.

The prompt flash of radiation from the nuclear detonation temporarily blinds hundreds of drivers along the major highways throughout the Los Angeles Basin, immediately causing a series of massive accidents and traffic jams throughout the Basin. Most severely affected are the main highways within ten miles of the detonation.

Destruction and damage of the power-grid nodes in the vicinity of the port cause widespread power outages throughout the Los Angeles area. In addition to the blast effects, the electromagnetic pulse (EMP) from the explosion destroys most electronic equipment within roughly a one-mile radius of ground zero.

Fires rapidly begin to burn around the port area and in the southwest corner of the city of Long Beach. Paper, wood, and other combustibles up to two miles from ground zero are ignited, setting many fires that quickly begin to burn out of control.

Early fallout is detected by firefighters two miles north of the port. High levels of radiation force them to retreat to the west.

Early estimates indicate that fallout will continue to be deposited downwind for approximately 24 to 48 hours, but just where the winds will blow the fallout remains uncertain. Initial projections based on surface wind speeds predict a fallout zone extending north of downtown Los Angeles, but experts warn that winds at a height of 20,000 feet are highly variable, based on prevailing wind patterns. (See Figure A.1.)

Aftermath: 24 Hours Later/June 7, 2009. Fires are now raging out of control at the port and fuel facilities, and in the surrounding waters from badly damaged ships and the tanker near the blast. First responders are not able (or are not allowed) to take action because of the high radiation levels from the early fallout. Visibility is severely limited by black smoke from the oil fires, which are also predicted to resuspend radioactive particles into the air.

Landlines and the few cell-phone towers that remained operational immediately after the initial blackout are quickly overloaded and are no longer operating. As a result, many people who were at work begin trying to head home to reunite with their families or pick up children from school.

Government Emergency Telephone Service and Wireless Priority Service remain effective on landline phones, where phones are available and in service.

The fallout zone has been carefully mapped by National Laboratory Teams, including local hot spots at which radiation levels are surprisingly high. The electric utilities have begun sending workers in to the cooler parts of the fallout zone for short periods to reset switches that affect delivery of electricity to the broader Los Angeles Basin.

Community officials are beginning to realize the magnitude of the fallout zone on low-income populations. The affected area comprises

Figure A.1
Fatalities as a Function of the Concentration of Fallout from the Nuclear Explosion

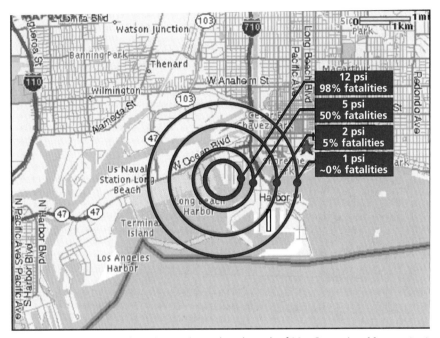

SOURCE: The MapQuest logo is a registered trademark of MapQuest, Inc. Map content © 2006 by MapQuest, Inc. and NavTeq Corporation. Used with permission. Map overlays added by authors.

communities in which languages other than English are spoken at home, including Spanish and many different Asian languages. Numerous hospitals and clinics to which low-income people turn for medical treatment have been contaminated and will be lost for years. This population, unable to flee far from the fallout zone except by publicly provided mass transit, continues to arrive at hospitals throughout the area for assistance and shelter.

A huge refugee problem has emerged. In addition to those who are displaced from the fallout zone, at least 4 million more people have self-evacuated out of fear of the fallout and other attacks. However, some, unable to leave the area because of impassable roads or lack of transportation, have remained in the Los Angeles region outside the fallout zone, staying with friends or squatting in parks, playing fields, or on beaches. Red Cross efforts to set up shelters are outstripped by demand.

Looting and crime have become major problems in the Los Angeles area. Desperate people seeking food cause some crime, but thugs looting abandoned residences and businesses and preying on refugees cause the most crime. Shootings have become commonplace.

The victims of the incident overwhelm hospitals. Shortages of personnel and supplies are reaching desperate levels. Mobile military medical units are helping to relieve the overload, but their efforts are not sufficient. State and federal officials are using such available large spaces as stadiums and gyms within a 50-mile radius as clinics.

Casualty estimates are that roughly 14,000 were killed at the blast site and in the heavy-fallout areas. Emergency clinics are processing thousands of additional cases of persons who will die shortly of radiation poisoning. Senior state public-health officials have made an estimate that an additional 50,000 to 70,000 individuals seriously ill with radiation poisoning will likely die at a declining rate over the next ten days.

Radiation in the immediate vicinity of Long Beach remains high.

Atlantic Tsunami (2010)

Background. Prompted by the Indian Ocean tsunami disaster of 2004, a global tsunami-alert system is fully operational by summer 2008. In the North Atlantic, the focus of concern is the active volcano, Mount Taburienta, on La Palma, an island in the Canary Islands. For several years, local geological monitoring stations have reported a steady rise of small earthquakes, early indications of a volcanic eruption. La Palma is a plausible source of a major tsunami that could menace the entire North Atlantic littoral.

By summer 2010, the U.S. Geological Survey and the National Oceanic and Atmospheric Administration (NOAA) have sent out a series of warnings about the prospect of a tsunami along the East Coast of the United States. These warnings prompt the Secretary of the Department of Homeland Security (DHS) to call a series of urgent meetings with the governors of the all states with Atlantic shorelines. The governors and DHS conclude a number of agreements that lead to a series of tsunami hotlines. Further, the public and media are informed of the potential threat. Several states, including Virginia, Maryland, and New York, test out their statewide capacity to get timely warnings to various coastal cities and towns. Simultaneously, the Secretary of Defense, with the concurrence of the Chairman of the Joint Chiefs of Staff (CJCS), orders the Combatant Commander of the Joint Forces Command (JFCOM) to prepare emergency plans, including the emergency evacuation of high-value naval and Coast Guard assets from vulnerable ports. Additionally, all military aviation assets are ordered to update the emergency evacuation plans that heretofore were primarily designed to deal with the threat of hurricanes.

Scenario: August 7, 2010. The volcano on the island starts a series of massive eruptions. Contrary to the expectation of several key European and American geologists, the volcano gives little warning of the massive eruption that occurs at 1630 local time, or 1130 EDT. Several media-coverage helicopters and light aircraft are flying within the vicinity of Mount Taburienta when it explodes. One helicopter is lost to falling debris. The surviving aircraft provide television images of both the initial eruption and rapid crumbling of a large portion of the island of La Palma into the Atlantic. Initial estimates are that more than 10 cubic miles of earth has shifted into the sea within a matter of minutes. Also within several minutes, both the U.S. Geological Survey and NOAA send out an emergency tsunami warning. Similar messages are sent out via like European and pan-African warning systems.

At 1245, the DHS alerts its East Coast subscribers to the tsunami warning system. The Combatant Commander of JFCOM orders the emergency evacuation of all major East Coast combatants capable of sailing within three-and-a-half hours. Early estimates suggest that the tsunami should begin hitting the East Coast around 1605 EDT. Every state from Maine to Florida orders an emergency evacuation of their coastal regions. This causes mass panic and many incidents of civil disorder in a number of coastal regions when it become apparent that transportation bottlenecks will not allow for quick evacuation. Most vulnerable are the barrier islands along Long Island, the North and South Carolina coast, and northeast Florida.

By 1500 EDT, all major warships that are not in drydock have successfully left Norfolk and Jacksonville, including three nuclear-powered carriers (CVNs) and three large-deck amphibious ships. Many of the ships are not with their full complement, and one of the CVNs, under maintenance, leaves with two of its four steam catapults non-operational. All Navy and U.S. Marine Corps (USMC) aircraft fly to airfields further inland.

At 1606, the first of two or three giant waves (the number of waves varies, depending on the specific area hit) break along the Outer Banks of North Carolina, with estimated heights of some 45 to 50 feet. During the course of the next hour, the entire East Coast is hit with tsunami waves of heights varying from 15 to 50 feet. Similar to the Indian Ocean Tsunami experience of 2004, rescuers will find many coastal communities with few casualties: Either the local population survived or is drowned. In other communities, however, many people are injured. Local media helicopters provide dramatic early pictures of entire barrier islands "disappeared" or cut into pieces.

By 1800, the Secretary of DHS reports to the President that early estimates of casualties indicate that more than 20,000 Americans have been killed, and more than a million have lost their homes. The most severe damage is the area from the North Carolina–Virginia border to roughly Charleston, South Carolina. Coastal towns in this area have been virtually obliterated. The Outer Banks of North Carolina have apparently suffered grievous damage, some islands being virtually washed away. Coastal highways and roads, bridges, and power grids have been demolished by the waves, leaving a number of coastal communities, particularly in the Carolinas, cut off and isolated from inland traffic. Extensive damage also extends into northern Florida, Georgia, Virginia, and Maryland. See Figure A.2. The docks at the Norfolk Naval Station and Newport News have suffered substantial damage. The president orders that all available military forces be used to provide immediate disaster relief where practicable.

Figure A.2
Southeastern U.S. Territory Affected by Atlantic Tsunami

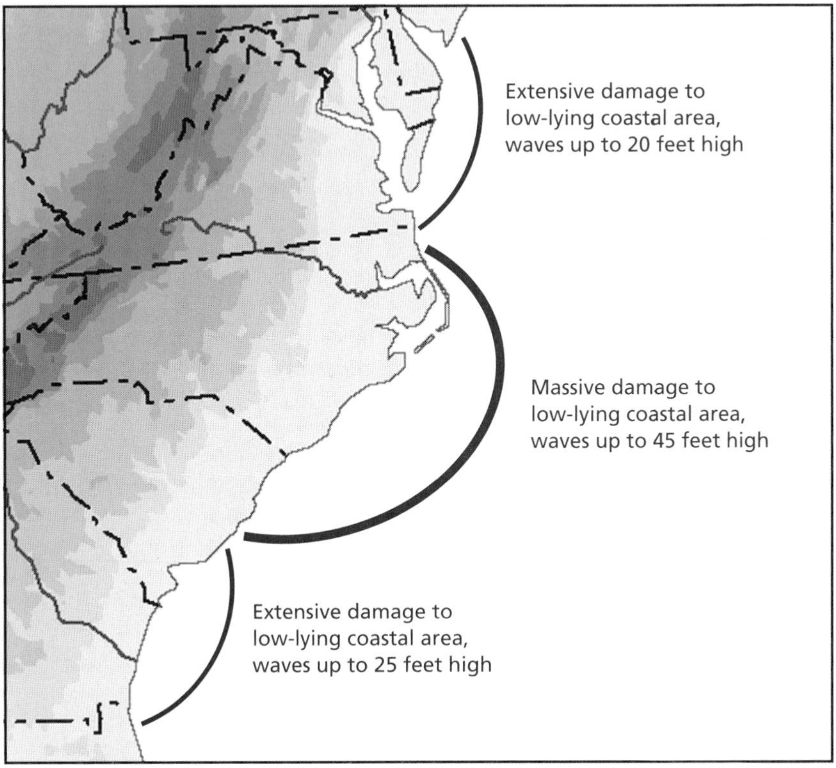

Extensive damage to
low-lying coastal area,
waves up to 20 feet high

Massive damage to
low-lying coastal area,
waves up to 45 feet high

Extensive damage to
low-lying coastal area,
waves up to 25 feet high

RAND *MG448-A.2*

The coastline along Portugal, western North Africa, and south-western Spain has suffered even more severe damage.

Massive Volcanic Eruption on the Island of Hawaii (2010)
Background. During 2007, the USN decides to home-base the CSG *Ronald Reagan* at Pearl Harbor. This is part of a larger restructuring of Pacific Command's (PACOM's) peacetime posture. As part of the repositioning and restructuring of the USMC's posture on Okinawa, the bulk of the 2nd Marine Division's aviation assets, especially its heli-

copters, is shifted to Oahu. Complementing this change is the home-porting of the ESG *Pelileu* at Pearl Harbor.

Throughout fall 2010, seismic activity on Hawaii, the "Big Island," picks up considerably. In turn, the Kilauea volcano begins a series of violent eruptions of "unprecedented" intensity.[1] These spectacular events during the course of September and October draw ever-larger crowds of sightseers to the national park. By late October, the intensity and frequency of the eruptions has alarmed the state and federal geological services, prompting them to warn the Governor of Hawaii that a mass evacuation of the Big Island might prove necessary. This forecast causes a major sensation in the national and international media and generates no little controversy. After much wrangling between NOAA, the state government, and the Department of the Interior, it is decided that Hawaii Volcano National Park should be evacuated of all nonessential scientific and law-enforcement personal by November 1, 2010. Evacuation plans of the surrounding communities are updated, and local residents are "encouraged" to find temporary lodging with family and friends elsewhere on the Big Island.

As part of the planning for any emergency response, the State of Hawaii's Office of Emergency Preparedness carries out a series of command exercises with the U.S. armed forces based on Oahu. In particular, the Navy notifies Hawaii's governor that all warships and aircraft on Oahu would be made available if the Big Island needed evacuation and/or other emergency assistance.

On November 5, the CBG *Ronald Reagan* returns from a major naval exercise off the Mariana Islands. The ESG *Pelileu* remains off the coast of Okinawa as part of a rapid South Korean reinforcement exercise staged in conjunction with elements of the Army's 2nd Division, which now is stationed on the island.

During the next several days, the eruptions at Kilauea fall nearly silent after a large portion of the crater floor appears to collapse. The

---

[1]   This scenario requires a geological change in Hawaii. The Kilauea volcano, unlike Mount St. Helens, for example, tends to vent pressure at regular intervals, which does not lead to an explosive eruption. This scenario assumes a major change in the volcano's internal configuration that could result in an explosive situation.

scientific community becomes increasingly alarmed by ever-clearer evidence that a massive plug is blocking further outgassing of the volcano. Further seismic evidence suggests that there is a major movement of magma below the volcano. By the evening of November 10, the Hawaiian governor is formally warned that "all residents within 30 miles of Mt. Kilauea should be evacuated." During the afternoon, the governor orders all residents, including those in the city of Hilo, to begin evacuation of the Big Island by the morning of November 11.

During the evening and early morning of November 10–11, the large community of scientists on the Big Island concludes that Kilauea will explode with great violence. All scientific personnel begin ground and air evacuation back to Hilo and Pahoa to the east, and to Kaalualu to the west. The Combatant Commander of PACOM formally orders all services on Oahu to make ready to provide emergency assistance to the State of Hawaii.

Scenario: November 11, 2010. At 0630, the state and federal geological science teams detect a massive shift of magma some ten miles below the crater. At 0701, Kilauea "blows its top." A vertical column of ash and rocks is shot up into the atmosphere to an estimated altitude of 90,000 feet during a series of explosions.

The shock waves of the explosions collapse man-made structures out to a distance of 10 miles. Windows are blown out in Hilo. The eruptions continue for another 40 minutes before subsiding. Much of the southeastern slope of Kilauea slides into the ocean with sufficient velocity to generate a major tsunami that will menace much of the coast of Central and South America later in the day. Fortunately, the islands north of the Big Island are largely protected from this event by the geometry of the tsunami event.

Unfortunately for the residents of the Hawaiian Islands, the prevailing wind is blowing toward the northwest at a brisk 20 knots. All animals and unprotected humans are killed out to the city limits of Hilo by the pyroclastic flow from a column of superheated rock and ash that later is estimated to contain more then 10 cubic miles of earth. Within hours, volcanic ash is raining down on Maui, Lanai, and Molokai. All airline flights out of Maui are cancelled. By late afternoon, the sky turns dark as ash begins to fall down on Oahu. The

Honolulu International Airport is closed at 1700 local time. Early and incomplete evidence suggests that more than 1,000 people may have been killed on the Big Island.

Earthquake Strikes San Francisco Bay Area (2009)
Background. During 2007, the U.S. Navy decides to home base the CSG *Ronald Reagan* at Pearl Harbor as part of a larger restructuring of PACOM's peacetime posture. As part of the repositioning and restructuring of the USMC's posture on Okinawa, most of the 2nd Marine Division's aviation assets, especially its helicopters, are shifted to Oahu. Complementing this change is the homeporting of the ESG *Pelileu* at Pearl Harbor.

Aside from the CSG *Abraham Lincoln* homeported in Japan, PACOM's three additional CSGs are based in San Diego. Two ESGs with associated large-deck amphibious ships are located at San Diego as well. After considerable debate, USS *Comfort* and USS *Mercy*, two single-purpose hospital ships, are retired for budget savings in 2008.

California and the "Not So Good Earth." By fall 2008, the U.S. geological community has become increasingly anxious about seismic activity up and down the San Andreas fault line. During Christmas Eve 2008, there is a Richter-7 earthquake under Lamont, just south of Bakersfield, which kills more than 100 people and injures several hundred others. Most casualties occur in several churches filled to capacity during Christmas Eve services. Aftershocks continue through the rest of December, with two registering Richter 6.

By February 2009, the state, federal, and academic geological communities have become even more alarmed. Evidence emerges that the recent Lamont quake and the aftershocks between Bakersfield and Fresno suggest that the deep elements of the San Andreas fault north of Los Baños toward Hayward, east of San Francisco Bay, has locked up. By late January, further evidence of major strain emerges south of San Jose, all the way down toward Salinas.

The Lamont earthquake and following seismic events prompt the governor of California to order his emergency services directorate to go on a high level of alert. In turn, the director of emergency services begins a series of round-robin consultations with various state and local emergency-response agencies. During this time, the combatant commander of the U.S. Northern Command (NORTHCOM) and the Secretary of DHS initiate a series of contingency-response exercises with the State of California. The combatant commander of PACOM is requested to allocate "appropriate" naval forces for a joint federal and state exercise starting on March 15.

The Korean Peninsula Heats Up/March 1, 2009. The PACOM combatant commander's interest in the NORTHCOM exercise is subordinated to the emergence of a severe military crisis in Northeast Asia. While enforcing an ever-more-stringent blockage of the DPRK's trafficking in missile- and nuclear-weapons-related technologies, a shoot-out at sea occurs between a North Korean vessel and a Japanese destroyer. The North Korean freighter blows up after a fierce engagement with the Japanese warship. The DNI warns the President that the North Korean regime might lash out. After a Cabinet-level National Security Council (NSC) discussion, the President orders the "leaning forward" of various military assets to Northeast Asia, including sailing the CSGs now in Japan and Hawaii. One of the CSGs based in San Diego prepares to depart on March 7. A second carrier is declared to be ready in 30 days.

Scenario: The Big One. At 1135 PST on March 21, 2009, the San Francisco Bay area is hit with a series of Richter-9 earthquakes. The epicenter of the first is underneath Cupertino, just south of Sunnyvale. The second, unprecedented Richter-9 seismic event occurs under Santa Cruz. Like a string of firecrackers, a series of Richter-7 and -6 shocks roll down the coast from Salinas to San Luis Obispo.

Within an hour, it has become apparent that a second "Big One" has occurred. Early estimates are that more than 4,000 people have been killed throughout California and that several tens of thousands are wounded. Both San Jose and San Francisco airports suffer severe damage and are closed.

Oakland International Airport, suffering minor damage, remains open to emergency traffic. Much of the road, rail, and electric-power infrastructure have been wrecked from San Jose up to Palo Alto. Early aerial video of Santa Cruz shows the town flattened and burning. State Highways 17, 1, and 101 have suffered severe damage. By 1430, it has become apparent that local fire departments are unable to cope with large, multiple fires burning in San Jose, Sunnyvale, Fremont, and Palo Alto.

The governor of California declares a statewide emergency and fully mobilizes the California National Guard. Within several hours after preliminary damage assessment, he calls the Secretary of Homeland Security for immediate federal assistance, including the early use of relevant elements of the U.S. armed forces.

The Navy is ordered to sail "the partially ready" CSG *Theodore Roosevelt* from San Diego when "equipped" for emergency homeland security support. One of the ready ESG with the *Essex* is ordered to sail north from San Diego as well.

Cuban Refugee Crisis (2006–2007)

Starting in late 2006, President of the Council of State and President of the Council of Ministers Fidel Castro Ruiz experiences a dramatic decline in health. In anticipation of his impending death and the scheduled elections in 2008, anti-Castro elements, both domestic and exiled, begin to position themselves for what they perceive to be a window of opportunity to challenge the National Assembly of People's Power and Fidel's heir apparent, First Vice President of the Council of State and First Vice President of the Council of Ministers General Raoul Castro Ruiz. With the aid of some Catholic clergy and some mid-level officers of the Revolutionary Army and Territorial Militia Troops, reformers quietly encourage popular support for an initiative that would establish a popular plebiscite and weaken the one-party system. Upon Fidel's death on August 17, 2007, the reformers openly advocate their initiative at the same time that Raoul comes to power.

In response to the open challenge, Raoul begins a purge of the military dissidents, and he brutally quells the resulting open, armed resistance. Although there is universal condemnation of Raoul's response from the international community, the Revolutionary Military continues to ferret out those who had supported the reformers. As a result, thousands begin to flee the island by boat to avoid death or imprisonment.

Most refugees flee to the north, toward the Florida Keys. To complicate matters, there are three tropical-storm systems in the Atlantic Ocean, any one of which could develop into a hurricane and threaten the Caribbean.

To avert an uncontrolled influx of Cuban refugees and a large-scale humanitarian disaster, the U.S. government begins diplomatic efforts to identify regional hosts to temporarily accept the refugees. At the same time, it orders the U.S. Southern Command (SOUTHCOM), NORTHCOM, and the Department of the Navy to assist the Department of Homeland Security in intercepting and providing humanitarian aid to the Cuban refugees.

The U.S. Navy is directed to send ships into the waters between Cuba and the southern United States to provide humanitarian assistance and to block unlawful entry of refugees into the United States. Carriers are included in the naval forces dispatched to the area. The Navy and Coast Guard will operate together to share the burden of this operation.

Bibliography

"Aircraft Carrier Firepower Demonstrated During Exercise," *Navy News This Week*, August 7, 1997.

Birkler, John, C. Richard Neu, and Glenn Kent, *Gaining New Military Capability: An Experiment in Concept Development,* Santa Monica, Calif.: RAND Corporation, MR-912-OSD, 1998.

"DCGS-N Budget Item Justification Sheet," *Department of the Navy FY 2006/FY 2007 President's Budget*, n.d. available at https://164.224.25.30/FY06.nsf/($reload)/85256F8A00606A1585256EFC00471ADA/$FILE/101OPN_FY06PB.pdf.

Jane's All the World's Aircraft 2003–2004. Available online at globalsecurity.org.

Jane's All the World's Ships 2001–2002. Available online at globalsecurity.org.

Lambeth, Benjamin S., "The New Face of Naval Strike Warfare: U.S. Carrier Air Operations and Capability Improvements Since Desert Storm," Santa Monica, Calif.: RAND Corporation, unpublished RAND research, 2005.

Larrabee, F. Stephen, John Gordon IV, and Peter A. Wilson, "The Right Stuff: Defense Planning Challenges for a New Century," *The National Interest*, Issue No. 77, Fall 2004.

McGuire, Bill, "The Future of Atlantic Tsunamis," *New Scientist,* October 22, 2005.

Morison, Samuel E., *The Battle of the Atlantic*, Boston: Little, Brown and Co., 1970.

————, *Operations in North African Waters*, Boston: Little, Brown and Co., 1968a.

————, *The Rising Sun in the Pacific*, Boston: Little, Brown and Co, 1968b.

Pendlow, Gregory W., and Donald Welzenbach, *The CIA and the U-2 Program, 1954–1974,* Washington, D.C.: Central Intelligence Agency, 1988.

Raytheon Technical Services, "Shared Reconnaissance Pod," n.d. Available online at http://www.raytheon.com/businesses/eps/pdfs/SHARP.pdf.

Toremans, Guy, "Combating the 21st Century Terrorist Threat," *Jane's Navy International*, February 23, 2005.

U.S. Navy, "Command, Control, Communications, Computers, and Intelligence (C4I)," in *Vision . . . Presence . . . Power,* 1998. Available online at http://www.chinfo.navy.mil/navpalib/policy/vision/vis98/vis-p11.html; last accessed March 9, 2006.

————, http://www.chinfo.navy.mil/navpalib/ships/carriers/histories.

————, *Naval Transformation Roadmap*, 2003. Available online at http://www.chinfo.navy.mil/navpalib/transformation/trans-toc.html; accessed February 19, 2006.